ETHEREUM FOR BEGINNERS

The Ultimate Guide to the World of Ethereum

(The Definitive Quick & Easy Blueprint to
Understand and Profit With Ethereum)

Emerita Malone

Published by Tomas Edwards

Emerita Malone

All Rights Reserved

Ethereum for Beginners: The Ultimate Guide to the World of Ethereum (The Definitive Quick & Easy Blueprint to Understand and Profit With Ethereum)

ISBN 978-1-990373-70-1

All rights reserved. No part of this guide may be reproduced in any form without permission in writing from the publisher except in the case of brief quotations embodied in critical articles or reviews.

Legal & Disclaimer

The information contained in this book is not designed to replace or take the place of any form of medicine or professional medical advice. The information in this book has been provided for educational and entertainment purposes only.

The information contained in this book has been compiled from sources deemed reliable, and it is accurate to the best of the Author's knowledge; however, the Author cannot guarantee its accuracy and validity and cannot be held liable for any errors or omissions. Changes are periodically made to this book. You must consult your doctor or get professional medical advice before using any of the suggested remedies, techniques, or information in this book.

Upon using the information contained in this book, you agree to hold harmless the Author from and against any damages, costs, and expenses, including any legal fees potentially resulting from the application of any of the information provided by this guide. This disclaimer applies to any damages or injury caused by the use and application, whether directly or indirectly, of any advice or information presented, whether for breach of contract, tort, negligence, personal injury, criminal intent, or under any other cause of action.

You agree to accept all risks of using the information presented inside this book. You need to consult a professional medical practitioner in order to ensure you are both able and healthy enough to participate in this program.

Table of Contents

INTRODUCTION 1

CHAPTER 1: ETHEREUM 101 3

CHAPTER 2: WHAT IS CRYPTOCURRENCY? 9

CHAPTER 3: WHAT IS ETHEREUM? 17

CHAPTER 4: WHAT CRYPTOCURRENCIES ARE GOOD TO INVEST IN? 25

CHAPTER 6: THE BIRTH OF ETHEREUM 31

CHAPTER 7: WHAT IS ETHEREUM 45

CHAPTER 8: WHAT IS ETHEREUM AND BLOCKCHAIN TECHNOLOGY? POTENTIAL BENEFITS OF ETHEREUM 58

CHAPTER 9: WHAT IS ETHEREUM? 72

CHAPTER 10: HOW ETHEREUM WORKS 81

CHAPTER 11: WHAT IS CRYPTOCURRENCY? 98

CHAPTER 12: MONEY, CRYPTOCURRENCY AND ETHEREUM 110

CHAPTER 13: THE RULES OF SERIOUS INVESTING 118

CHAPTER 14: WHAT IS ETHEREUM? 128

CHAPTER 15: GAIN WITH ETHEREUM 133

CHAPTER 16: WHAT YOU NEED TO GET STARTED 146

CHAPTER 17: ETHEREUM'S FUTURE 153

CHAPTER 18: ETHEREUM AND BITCOIN – A COMPARISON 161

CHAPTER 19: HOW TO BUY, SELL, AND STORE ETHER 167

CHAPTER 20: USEFUL TIPS 185

CHAPTER 21: TOP TIPS TO MAKE YOUR ETHEREUM EXPERIENCE BETTER 187

CHAPTER 22: THE FUTURE OF ETHEREUM 194

CONCLUSION 198

Introduction

The following chapters will discuss the technology behind Ethereum.

You may not know much about Ethereum, but you may know about bitcoin. Ethereum is not going to be much different from bitcoin since they are both going to be blockchain applications. However, you will have more options when it comes to investing in the platform. This book will give you all of the information that you need to make an informed decision on if you want to invest with Ethereum or not.

Ethereum is a platform that will allow you to trade a digital currency that is also going to make it to where you can write outsmart contracts so that you are not having to worry about spending a lot of money.

One of Ethereum's biggest goals is to save its users money, and from what you will notice in this book, you will see that they will be doing just that. You are not only going to be saving money, but you will be getting rid of outside variables that could end up messing up your investments.

Another great thing that you will notice is that Ethereum is not going to require a lot of money to invest, unlike the stock market.

There are plenty of books on this subject on the market, thanks again for choosing this one! Every effort was made to ensure it is full of as much useful information as possible; please enjoy!

Chapter 1: Ethereum 101

We define Ethereum as the blockchain platform on which the currency known as Ether is exchanged. The blockchain is open to all for development such that anyone can design programmable functions for executing transactions and trades. That makes Ether and Ethereum a very decentralized system lowering inherent risks and improving the user's functional control over the currency's possibilities.

While Bitcoin was developed in early 2008 and released to the public in 2009, Ethereum came four years later in 2013 and was presented by Vitalik Buterin. It was not until the following year that the infrastructure housing the decentralized blockchain was put into use, and still another year in 2015 until the public blockchain was released. The Russian creator developed the Ether system in Canada while actively involved in further developing Bitcoin. However, it was apparent that benefits could be made to the concept of Bitcoin, but would require its own blockchain to provide a complete flexibility in development and creation of applications. Thus, Buterin opted to use a Turing complete programming language. His invention awarded him the World Technology Award for the cryptocurrency's invention. You may be asking what Turing complete means.

Well, you don't need to understand a large amount of math or computer science to be able to comprehend why Turing complete is an advancement for cryptocurrencies. What you do need to know is that Bitcoin, which is operated in a computer language that is **not** Turing complete simply cannot express certain types of actions or contract requirements that would be essential to make projects like Ethereum work.

What had been happening previously was for every specific application that Bitcoin couldn't meet, people would create a new, alternative type of cryptocurrency or altcoin, to produce the results needed for each specific application Bitcoin wasn't able to handle. In one such example,

NameCoin had to be developed to register domain names that would be resistant to any type of censorship. This is because Bitcoin couldn't access the type of language needed to get into registering domain names. Another important example is Litecoin which is basically the same as Bitcoin, except it's all open source. Due to the scripting and hashing of Litecoin, however, blocks can be generated at about 1/4th the time compared to Bitcoin, making mining more efficient. If you're interested in all the CCs, definitely research Litecoin.

In a more broad and fundamental approach, Ethereum's foundation is still the Bitcoin blockchain type of technology, but the platform now has a Turing complete programming language built into it. That means the platform can now cross-talk with any type of application you desire, and thus the building of apps (Dapps) and contracts becomes a paramount strength in Ethereum's framework.

Although Ethereum is decentralized, there is a major company and foundation that pilot the paradigm of the platform. Ethereum Switzerland GmbH was the flagship program developer of all things Ethereum at its launch on July 30, 2015.

Additionally, the Ethereum Foundation helps produce and maintain the collective use of Ether coin and is also based in Switzerland. Vitalik Buterin is the lead for the research team at Ethereum Foundation, working alongside others to produce the next vision of Ethereum and its decentralized apps (Dapps).

It only took one year from public release for the value of Ethereum to reach $1 billion USD and begin to give Bitcoin a run for its money. The simple fact is that Ethereum can do a variety of new things that Bitcoin cannot.

Ether and Value

The value token of Ethereum is the Ether or Ether coin. Comparative to Bitcoin, it is exchanged on blockchain in transactions, and carries with it all the benefits of micro transactions (being able to send any fractional value of a single, whole Ether). The Ether currency will be held in your wallet and can be used in the Ethereum network. There is no federally-backed monetary value of Ether or any cryptocurrency which actually brings up an interesting point about monetary value in its most fundamental aspect; monetary value is completely dependent on people's agreement of it.

When people speak about how paper money used to be backed by the value of gold, i.e. every dollar had an exact exchange rate for gold, people felt safe. When the dollar transitioned from commodity money (representative of precious metal) to fiat money (government-assigned value via federal decree), people became afraid. What did money even mean anymore if gold didn't back it? On further analysis, gold itself has very little intrinsic worth and is actually completely dependent on people's choice of it being an aesthetically pleasing metal! There are properties to gold that make it valuable, but to act as a currency, all people within a given market must agree on the value. Thus, the notion of value between our fiat and commodity currencies used around the globe today is fundamentally no different than assigning a value to cryptocurrencies that ultimately bear no physical form.

There is an obvious fundamental difference between any physical currency and that of Ethereum: you need the internet. Since Ethereum is young and at the time of writing this book barely over a year old, it's acceptance is much more limited than that of the dollar or even Bitcoin. Thus, describing Ethereum as a speculative investment is an understatement. However, as I will explain, making the contracts, to make Ether, should benefit in returns with the more effort you put into it.

Chapter 2: What Is Cryptocurrency?

A cryptocurrency can be defined as a digital currency that is used online. Cryptocurrencies use cryptography technology to secure the currency and all transactions. The currency is a virtual currency which does not exist in any tangible form. Transactions are anonymous and secured by the technology used.

Cryptocurrencies operate on a distributed ledger system enabling all users to view transactions. There is no central control and no government or authority oversees the currency. This decentralized nature of cryptocurrency is one of its strongest attributes.

Cryptography technology emerged from the desire for secure communication. Modern cryptography has embraced elements of computer science and mathematical theory to provide a secure form of communication. Using cryptograph technology, information is encrypted and sent in a secure and private way that is almost impossible to hack. The technology is used to send information securely and for data security and authentication. It is also used for secure funds and communications online.

Some years prior to 2009, there had been plans to create a decentralized currency that used a distributed network and cryptograph technology. Finally in 2009, the first such currency came into existence. This is Bitcoin which to date is the most popular cryptocurrency in the world today.

One of the most outstanding features of cryptocurrency is its organic nature. Cryptocurrencies are not controlled by governments or other centralized entities. This ensures they are free from government manipulation or interference.

Cryptocurrencies are not printed or minted like coins and are not backed by tangible assets such as gold. Instead, they come into existence through a process known as mining. Miners spend their time solving complex mathematical puzzles and when they do this cryptocurrencies come into existence.

How do cryptocurrencies work?

Cryptocurrencies are anonymous in nature. Users do not need to have their names attached to transactions and there is no need for bank accounts. The technology used is decentralized so users can send and receive money or make payments via a distributed public ledger which is known as the blockchain.

The blockchain is a public ledger where all transactions are recorded. It is not a static ledger because it is always being updated. All users have access to the blockchain and can view a history of all transactions. The distributed ledger network is also a peer-to-peer system where users can transact directly amongst themselves without the need for a third party.

Anyone from any part of the world is free to use cryptocurrencies. There is no limitation based on gender, race, creed, location or nationality. There are online platforms and exchanges where cryptocurrencies can be purchased or exchanged for other currencies.

Each account or user is linked to all others via complex mathematical equations. This is done to ensure that each transaction is legitimate and accurate. The job of confirming transactions, legitimizing them and adding them to the blockchain is done by miners. Miners are usually computer programmers and software engineers. They are rewarded for their work through the transaction fees charged and the currency they mine.

Virtual currencies make transfer of cash from one party to another convenient, fast and easy. To transfer cryptocurrency from one user to another requires the use of tow distinct keys. Since virtual currencies are not tangible, they are represented by a string of alphanumeric characters referred to as keys. One key is the private key while the other is a public key. Keys are also encrypted in this manner to secure them.

When funds are transferred from one user to another, there is no transaction fee. Sometimes there is a nominal fee charged.

This fee is absolutely minimal and is mostly used to reward miners, the experts who confirm transactions and add them to the blockchain.

Challenges

There are challenges pertaining to cryptocurrencies. One of the major challenges stems from the digital nature of these currencies. Since they are virtual and only exist in cyberspace, users could lose all their funds through a computer crash if there was no backup. There are firms that have lost millions of dollars worth of cryptocurrencies due to a computer crash. Backing up data running on cryptocurrency systems is very important.

Volatility

Cryptocurrencies are very volatile. The price or value often depends on demand and supply. Investors may lose a lot of money if they purchase large quantities of cryptocurrencies only for the prices to dip a short while later. The volatility of cryptocurrencies is largely due to speculative nature of users. Many users buy the digital currencies and at the slightest price difference rush to sell in the hope of making a profit. To drastically reduce this, users should learn to buy the currency and hold onto it for a while.

Hacking

Hacking is a problem that is not uncommon with cryptocurrencies. There have been a couple of major attacks that have touched on popular cryptocurrencies. For instance, Bitcoin, the most popular currency has been hacked and funds stolen over 40 times. A couple of these thefts involved losses worth over $1 million. Fortunately, the weaknesses were discovered and fixed.

Users are advised to take precaution and protect their investments. For instance, they can use a strong password, adopt multiple signatures and store cryptocurrencies offline. Such simple steps can secure cryptocurrencies and keep them safe from hackers and online thieves. Even with all these challenges, plenty of observers are of the opinion that cryptocurrencies are here to stay and will, in the future, be widely accepted globally.

These currencies preserve value, are not under any government control, are easier to move between countries and facilitate trade.

Cryptocurrency summary

In general, there are over 700 cryptocurrencies in the world today. All these digital currencies were developed after Bitcoin, the initial cryptocurrency. Cryptocurrencies other than Bitcoin are collectively referred to as altcoins.

There are exchanges where investors and ordinary users can buy cryptocurrencies. Before investing in digital currency, users are advised to conduct due diligence. Many of these are unstable and some even fail completely, causing investors to lose funds.

Instead of speculation, investors should undertake research, patience and an abundance of caution.

Some of the most successful altcoins in the market today include Ethereum and Litecoin. It is advisable to remember that there are users out there waiting for new investors to take their money.

Chapter 3: What Is Ethereum?

It will be easier to understand the nature of Ethereum if you have background knowledge on how the Internet works.

At present, our personal profile, financial information, and passwords are mostly stored in servers and data storage controlled by big players like Google, Facebook, and Amazon. Even the articles and blogs you are reading online are stored on a cloud owned by an organization that requires fees for keeping all the data.

This arrangement has numerous advantages as these organizations employ teams of data professionals to assist in keeping and securing information and help in eliminating the costs that usually come with hosting and uptime.

But there are also loopholes with these advantages. Cyber criminals, even government agencies, can have access to your data without your permission, by hacking or controlling a third-party service.

Your personal information could be leaked or even modified.

The founder of Apache Web Server, Brian Behlendorf has gone so far as calling this centralized setup the "original sin" of the Internet. Advocates like him insist that the Internet should be decentralized. They highlight the risks of having a centralized design of the World Wide Web. Well, a movement has emerged around utilizing new tools such as the blockchain technology, to help attain this objective.

One of the recent technologies that take part in this splintered movement is Ethereum. Unlike Bitcoin, which aims to replace online banking, the main objective of Ethereum is to use blockchain to disrupt online third-party services.

The Global Computer

Ethereum's goal is to be a "global computer" that would not stick to the centralized user-server setup. There are also people who argue that Ethereum will democratize the current system.

With Ethereum, nodes will replace the conventional clouds and servers. These communication points are controlled by groups of volunteers located around the world.

The objective is that Ethereum would perform the same functionality to users anywhere across the globe, allowing people to compete in providing services on top of this infrastructure.

Browsing through an application store, for instance, you will find different types of applications that can be used in every aspect of life. These applications depend on a third-party service to keep your financial information, transaction data, and other personal information in their respective clouds or servers.

If everything works according to the vision, Ethereum will allow the owner of these applications to take control of the data as well as have creative rights. The point is that third parties can no longer have access to your data, and that one entity can no longer curate or censor your applications. Again, only the owner can input changes, not any entity.

In the proposed theory, it integrates the control that users had over their data in the past with accessible insights that we're used to in the digital era. Every time you input or delete notes, every communication point on the web makes the change.

Somehow the concept has been mixed with skepticism.

Even though the applications don't seem impossible, it's uncertain which apps will actually be helpful, secure, or can be quantified – and if they will be as efficient to use as the applications we have today.

With blockchain technology, Ethereum became an open source tool that gives developers authority to create over a decentralized app. With its highest hash rate value, which is 3 TeraHash, it has a huge infrastructure. Vitalik Buterin with the help of his co-developers, Charles Hoskinson, Mihai Alisie, and Anthony Di Lorio, created Ethereum for high-grade graphics processing units.

Because Ethereum is working on blockchain networks, it also comes with the conveniences of decentralized networks such as:

- Inability of other entities to input changes on data.
- Prevention of fraudulent activities caused by cyber criminals or hackers.
- Zero percent chance of applications to be turned off or crash.

The distribution of Ethereum was made possible through a public blockchain network, which is in the form of Initial Coin Offering (ICO). In ICO, more or less 35,000 Bitcoins were traded for about 60 million. This allowed an estimated amount of $ 14 million to be raised, which comprised 14% of stocks.

Moreover, the distribution of Ethereum continues to work through an ICO, in which it works at the same time with the distribution of Ether. ETH or Ether is known as the Ethereum platform's cryptocurrency token.

In Ethereum, a blockchain introduces blocks of various scalability. At the same time, there are different kinds of user accounts that raise with 22 byte addresses. There are two kinds of accounts – the external account which is the account with private keys, and the other one is the contract account which is the account with contract codes.

The authority that controls them is one of the major distinctions between the external and contract accounts.

Contract accounts are controlled by internal codes. While they can utilize contract accounts, people need external accounts to enable contract accounts. Moreover, contract accounts are permitted to execute transactions if ordered by external accounts.

Therefore, unless triggered by external accounts, these contract accounts cannot execute their usual operations such as Random Number Generation and API calls. On the other side, the control for external accounts is handed to human users since these users can control private keys, which consequently return control to external accounts.

Smart Contracts

Smart Contracts (or the use of scripting functionality) is one of the most significant features of Ethereum platform. This allows users to produce tokens that are suitable with wallets and exchanges under a standard coin API (application programming interface).

Moreover, this shows that it can speed up the exchange of money, stocks, or property. And thereby, with the use of smart contracts, we can count on a self-operating program that facilitates orders immediately once conditions are approved. Since their works are executed through a blockchain network, they operate according to how it was programmed. They execute without the possibility of being interfered by other entities, and without the chance of censorship or downtime.

Here's an overview of a smart contract's operations:

- Initially, a code from an option contract is given into the blockchain network. Although the contract serves as the public ledger, this code is given with the security from the concealment of both parties.

- A smart contract is completed once a period expires and strike prices has been reached.
- Finally, since a smart contract was executed, regulators will start monitoring the activities of the market by utilizing the blockchain network. Nevertheless, all throughout the monitoring, regulators are required to secure the privacy of both parties.

Chapter 4: What Cryptocurrencies Are Good To Invest In?

This year the value of Bitcoin has soared, even past one gold-ounce. There are also new cryptocurrencies on the market, which is even more surprising which brings cryptocoins' worth up to more than one hundred billion. On the other hand, the longer term cryptocurrency-outlook is somewhat of a blur. There are squabbles of lack of progress among its core developers which make it less alluring as a long term investment and as a system of payment.

Ethereum

Vitalik Buterin, superstar programmer thought up Ethereum, which can do everything Bitcoin is able to do. However its purpose, primarily, is to be a platform to build decentralized applications. The blockchains are where the differences between the two lie. Basically, the blockchain of Bitcoin records a contract-type, one that states whether funds have been moved from one digital address to another address. However, there is significant expansion with Ethereum as it has a more advanced language script and has a more complex, broader scope of applications.

Projects began to sprout on top of Ethereum when developers began noticing its better qualities. Through token crowd sales, some have even raised dollars by the millions and this is still an ongoing trend even to this day. The fact that you can build wonderful things on the Ethereum platform makes it almost like the internet itself. This caused a skyrocketing in the price so if you purchased a hundred dollars' worth of Ethereum early this year, it would not be valued at almost $3000.

Bitcoin

Still the most popular, Bitcoin is the cryptocurrency that started all of it. It is currently the biggest market cap at around $41 billion and has been around for the past 8 years. Around the world, Bitcoin has been widely used and so far there is no easy to exploit weakness in the method it works. Both as a payment system and as a stored value, Bitcoin enables users to easily receive and send bitcoins. The concept of the blockchain is the basis in which Bitcoin is based. It is necessary to understand the blockchain concept to get a sense of what the cryptocurrencies are all about.

To put it simply, blockchain is a database distribution that stores every network transaction as a data-chunk called a "block." Each user has blockchain copies so when Alice sends 1 bitcoin to Mark, every person on the network knows it.

Litecoin

One alternative to Bitcoin, Litecoin attempts to resolve many of the issues that hold Bitcoin down. It is not quite as resilient as Ethereum with its value derived mostly from adoption of solid users. It pays to note that Charlie Lee, ex-Googler leads Litecoin. He is also practicing transparency with what he is doing with Litecoin and is quite active on Twitter.

Litecoin was Bitcoin's second fiddle for quite some time but things started changing early in the year of 2017. First, Litecoin was adopted by Coinbase along with Ethereum and Bitcoin. Next, Litecoin fixed the Bitcoin issue by adopting the technology of Segregated Witness. This gave it the capacity to lower transaction fees and do more. The deciding factor, however, was when Charlie Lee decided to put his sole focus on Litecoin and even left Coinbase, where're he was the Engineering Director, just for Litecoin. Due to this, the price of Litecoin rose in the last couple of months with its strongest factor being the fact that it could be a true alternative to
Bitcoin.

Monero

Monero aims to solve the issue of anonymous transactions. Even if this currency was perceived to be a method of laundering money, Monero aims to change this. Basically, the difference between Monero and Bitcoin is that Bitcoin features a transparent blockchain with every transaction public and recorded. With Bitcoin, anyone can see how and where the money was moved. There is some somewhat imperfect anonymity on Bitcoin, however. In contrast, Monero has an opaque rather than transparent transaction method. No one is quite sold on this method but since some folks love privacy for whatever purpose, Monero is here to stay.

Zcash

Not unlike Monero, Zcash also aims to solve the issues that Bitcoin has. The difference is that rather than being completely transparent, Monero is only partially public in its blockchain style. Zcash also aims to solve the problem of anonymous transactions. After all, no every person loves showing how much money they actually spent on memorabilia by Star Wars. Thus, the conclusion is that this type of cryptocoin really does have an audience and a demand, although it's hard to point out which cryptocurrency that focuses on privacy will eventually come out on top of the pile.

Bancor

Also known as a "smart token," Bancor is the new generation standard of cryptocurrencies which can hold more than one token on reserve. Basically, Bancor attempts to make it easy to trade, manage and create tokens by increasing their level of liquidity and letting them have a market price that is automated. At the moment, Bancor has a product on the front-end that includes a wallet and the creation of a smart token. There are also features in the community such as stats, profiles and discussions. In a nutshell, the protocol of Bancor enables the discovery of a price built-in as well as a mechanism for liquidity for smart contractual tokens through a mechanism of innovative reserve. Through smart contract, you can instantly liquidate or purchase any of the tokens within the reserve of Bancor. With Bancor, you can create new cryptocoins with ease. Now who wouldn't want that?

Chapter 6: The Birth Of Ethereum

Before we can fully understand what Ethereum really is, we need to go back to the beginning and understand what money is and its original purpose. For centuries, money has been used as a means of transferring value from one person to another. It was a very effective means of keeping track of who owned what and was the next evolutionary step from the old barter system that many cultures once used.

The reason the barter system eventually fell into disuse was its impracticality. There were elements of the system that failed to take certain factors into effect. First, for an exchange to take place, the other party needed to specifically want the goods that others had to offer. For example, if you wanted to trade a plow for a cow, the other person had to want your cow. It also didn't take into account that each individual could set up his own value system, which wasn't likely to match with someone what another person believed was its true worth.

The introduction of money changed all of that. By having one central authority that determined a set value of the local currency, it served as a sort of equalizer so that value was a bit more consistent throughout a particular civilization. Once money became the common form of currency, it was monitored, issued, and managed by the governments.

This is the system we have all become very comfortable with and for thousands of years have used it with a great deal of success. Still, that doesn't mean that money itself didn't have any flaws of its own. Several problems resulted from the centralization of money. One of the main issues many faced was the need to use a central authority to transfer value from one person to the next. By using these central figures, users were often exposed to high fees, loss of privacy, and other governmental controls. In short, users often lost the ability to control and manage the funds that they worked for.

To understand this better, think about how you get paid for the work you do. In most cases, you go to work for the assigned pay period, but when it comes time to get paid, your employer does not just give you the money for the services provided. Instead, it is sent to a bank that is governed and regulated by the government; you receive a piece of paper that says exactly how much you earned. For you to access that money, you must first go to the bank and present some sort of government-issued identification or use a bank-issued card in an ATM machine before the money is issued to you. Before that happens, however, the government takes his cut, and the bank also charges its fees. All of that happens before you see your first dime.

If you want to pay the bill, then the same process happens. You must either ask the bank to send the specified amount to the merchant to cover your debt. Again, the merchant doesn't actually get the money directly unless the bank chooses to release the funds. At any rate, you get the picture.

We may think we are in control of our money, but it is really in the hands of uninterested third parties who also want to charge phenomenal sums of money for the unwanted privilege of handling our money.

This kind of management, while proven to be more effective than the old barter system, has had its own share of problems. As more and more third party individuals start to get greedy, charging even higher fees, people became increasingly more desperate to find other options for managing their money.

Many saw the problem stemmed from the use of a centralized system to manage the money. Even when they used digital currency, as they often did with services like PayPal, Skrill, or Google Wallet, there was always a high level of risk. The exposure to hacking and other cybercrimes were very real, and there was still the lack of control for many who are forced to rely on the system. Their privacy and their security were often compromised.

The idea for a change came literally from the world-wide-web, which for the first time in our history, made it possible for everyone to access the same information whenever they wanted it. It laid the foundation for economic giants like Amazon, eBay, and Apple to build multi-billion dollar corporations.

Once people got a taste of more control, felt what it was like to manage their own funds, it was just a matter of time before it would evolve into a means of managing the economics of society. This was to be done through the introduction of Blockchain Technology. It allowed for the control of currency to be put back into the hands of those who actually own it, worked for it, and used it. It made it possible for people to exchange value without the constant overseeing eye of Big Brother watching. Blockchain Technology allowed people to reclaim and control what they were working for all along.

What Exactly is the Blockchain?

In the simplest of terms, a Blockchain is merely a digital ledger similar to the one used at your bank. It is a record of what values exist and who owns them. The difference between the ledgers that the banks and governments maintain is a matter of who controls it and who has access to it. A centralized ledger owned by the banks is available to only a select group of people whereas the Blockchain ledger is open to the public. More specifically, a Blockchain is actually open-source software, which means that anyone on the system can observe transactions as they happen. Nothing on the Blockchain is done in secret.

This raises a common question: if the information on the Blockchain is public, how is it that it can protect someone's privacy? How can it be public but the users remain anonymous at the same time? The answer is where cryptocurrency comes in. The method of digitally encrypting data related to each transaction makes all of this possible.

Once a transaction enters the system, it is encrypted so that privacy is not only maintained, but the details of the transaction are virtually unalterable. At the same time, every detail of the transaction is placed in a block, which is then delivered to every computer in the system so that no single entity is in control of its contents. By taking the role of the middleman in all currency transactions and placing it in the hands of an automated third party, it basically boosts security in a number of ways. How exactly does this work?

In our traditional system, the banks would control the ledger and would monitor when transactions occur and maintain that information on a single server. The Blockchain, however, takes the role of the bank and puts the verification process in the hands of a group of people who manage the automation of the system. The identity and any other personal data related to those involved in the process are encrypted leaving only the total amount of the transaction open to the public. Because of this encryption coding, it is the most secure way of transferring value ever used by the masses. So, while the information is public and easily accessible to anyone using the network, the details of the transaction, personal identity, and other private information is thoroughly encrypted in forms that are virtually impossible for cybercriminals to get at.

Why It Was Needed

The Blockchain has proven to be a major breakthrough for many. Even if you have a limited background in finance, it is easy to understand how this could be of significance to many people. When Bitcoin first introduced it, it took a while before the impact of its value gained attention.

Bitcoin was used to offer an alternative form of currency, one that would always exist purely in the digital world. It is a system that allowed users to transfer value from one person to the next bypassing the current system. While banks and other financial institutions, could issue you physical money in exchange for digital currency, Bitcoin could never be used anywhere except through the Internet.

Because of the Blockchain, users were able to easily send currency anywhere around the world without the aid of a third party. If you wanted to make a purchase, you could pay using Bitcoin in much the same way as you would pay for an online purchase for the same item. They could also perform transactions much faster because there was no need for validation from a centralized third party.

This process makes it possible for a form of peer-to-peer transactions to be made merely on the basis of trust between the two parties involved. While it is similar to a straight cash transaction, where two people agree on the transfer of ownership of an item, it differs in the sense that the details of that transaction are recorded in perpetuity on the Blockchain. The "trust" therefore, is based not just on a verbal agreement but also in the evidence of that transaction that is recorded in encrypted digital format.

Bitcoin, therefore, was a whole different animal. It was similar in form to other digital currencies, but by using the

Blockchain format, it was no longer regulated by the governing bodies of whatever financial institution managed it. It also differed in the sense that no governmental institution would determine its value. Its ledger is totally maintained by the users of the network, and they are the ones that give it value.

It wasn't long before people began to perceive the true value of the Blockchain, which for a time was limited to the process of exchanging currency. People began to realize that the Blockchain could be utilized in others ways that extended far beyond that of finances.

It was from this type of innovative thinking that Ethereum was born. By taking Blockchain technology and using it in an entirely different way, users soon were able to create decentralized programs with far more complexity than Bitcoin.

What is Ethereum?

The creator of Ethereum, Vitalik Buterin, envisioned an open-source computer program that could be used as a platform for a wide range of programming capabilities. This opened the door for all sorts of innovative ideas to be put into practice without fear of your property being stolen. In essence, Ethereum's adaption of Blockchain technology created not just a platform of peer-to-peer exchange of currency but as a peer-to-peer transfer of computer programming. You can liken it to a collaboration of efforts in creating a single document; several people may have had some role in its final completion. The only difference is that Ethereum is still evolving. With each new addition to the code, new changes are being born.

The entire structure of Ethereum is exceedingly more complicated than Bitcoin's basic format of currency exchange. The transactions performed on Ethereum's network go much further than just an exchange of value.

It's Purpose

Two features make Ethereum stand out from other forms of cryptocurrency, the ability to create Smart Contracts and Distributed Applications (Dapps). These are the key elements of the Ethereum network. With numerous new applications already in use and even more still to be developed, these two features give us a clear picture of Ethereum's main purposes and serve as an impetus to its gaining popularity.

To put it in the most basic of terms, Ethereum is a decentralized platform used to run smart contracts and distributed applications without third party involvement. The network has zero risks of censorship, fraud, or any other type of interference that could prevent it from performing its tasks. With these smart contracts, a user can add just about anything to the Blockchain and know that it is secure and the terms of any agreement will be met without disruption.

Its coin, Ether is used to run the network and even to give value to work being done, but its applications can serve as a means to trade all manner of things that go far beyond currency. Through this platform, partnerships are being born every day. Most notably is a partnership between Microsoft and ConsenSys. Together, with the use of a smart contract, both their clients and developers can offer a single click, cloud-based Blockchain environment that can be utilized in all sorts of businesses.

It serves as a higher form of Blockchain technology that is similar in basic format to Bitcoin but does not share enough similarities to compete with its forerunner. With Ethereum, users can build and use their own decentralized applications meant to run with Blockchain technology in a way that is both adaptable to all sorts of business interests and flexible enough to be used by anyone.

In effect, while the two utilize the same Blockchain, they have two entirely different purposes. Bitcoin's main purpose is to provide an alternative means of exchanging value while Ethereum's main goal is to provide an alternative means of transacting all other forms of business.

How it Differs From Bitcoin

On the surface, the differences between the two are pretty clear. Ethereum is not and was never meant to be a form of purchasing power like Bitcoin, but the main differences lie it their technology. For example, the programming language is very different and unique to each coin. Other differences are found in the speed of transactions (Ethereum's transactions are completed in seconds compared to Bitcoin's minutes)

The most publicized difference is probably that Ethereum was never meant as a medium of payment but rather a means of creating peer-to-peer contracts and applications to facilitate the work developers need to do in creating DApps.

Chapter 7: What Is Ethereum

Ethereum could be called a digital coin in the digital world. Slowly approaching the original digital currency - bit, quickly became the second most valuable method in the new payment system. Since Ethereum first appeared in 2015, its value has increased only in the previous year, and now it is at almost $300. Ether is the second most valuable digital currency based on Ethereum technology, first described by the programmer. It designed the ether as a currency that will strengthen Bitcoin, which, like Bitcoin, will serve as a decentralized payment method, its own cryptographic a currency that allows users to make anonymous payments online without the need for a bank account and any third party. Transactions are stored in a decentralized system and can be seen by everyone within the network.

This currency claims to have certain advantages over Bitcoin. It can make it more useful. Namely, the ether allows the so-called blocks, evidence of crypto transactions, to be created much faster than in Bitcoin. It would be far more efficient in online trade. The biggest advantage of ether is that technology also allows computer applications to function within the network. It is important because of security of digital wallets. For example, if you upload some files to Dropbox, you believe that this platform will be responsible for taking care of them. When it comes to some decentralized storage network, your trust applies to all those who use it, and have any interest in is maintained so that some kind of control should exist. Ethereum has several applications installed, and the service also uses a startup company, which exchanges ether or other digital currencies for special tokens with they gain access to the service.

The value of the ether grew much slower than in the case of Bitcoin, and for the first 18 months, it was $10. Then, in March 2017, it began to grow and rose to $395 in June. Then dropped to $155, and recently climbed to nearly $300. Growth also provided an interest in Bitcoin, which awakened interest in all other digital currencies.

It is estimated that there are about 5.3 million digital wallets in which the ether is stored, since the number had risen since May 2016, when it was about 1.6 million. This currency, like other crypts, is known for allowing anonymous transactions and keeps it from fraud and stealing. Users can make large-scale transactions without paying a fee, since a third party, bank-filled, does not exist. It is possible to spend it in a large number of online stores, and with the help of credit cards for crypts. It can also be stored in a digital wallet until the currency value is increased, making the users significantly richer.

The internet should always be decentralized; in that connection, a movement has been used that has used new tools, such as blockchain technology, to achieve this goal.

Ethereum is one of the newest technologies to join this movement.

Ethereum aims to use a blockchain to replace these third-party internet users – those store data, transfer a mortgage and take into account complex financial instruments.

In short, the Ethereum wants to be a world computer that would decentralize - some would say, democratize - the existing client-server model. They replace thousands of so-called knots by volunteers from around the world by ether, servers, and clouds forming a world computer. The vision is that the Ethereum enables this same functionality to people anywhere around the world, enabling them to compete for providing services based on this infrastructure provided by ETH.

Ethereum needs to restore the control of data in these types of services to the owner and the creative rights of its author. It is a matter that an entity will no longer have control over your notes and that no one can suddenly ban the application, temporarily taking all of its notebooks. Only users can make changes, not any other entities. The entire process within the Ethereum network is called the Smart Contract. Theoretically, it's a combination of controlling your own information that people had in the past with easily accessible information that we used to get used to in the digital age. Every time you save changes or add or delete notes, each node in the network makes a change.

Ethereum is an open blockchain platform (decentralized chain of records) that allows anyone to build and use decentralized applications (called smart contracts) that implement blockchain technology. Ethereum is often referred to as a world-class computer because computer operations are executed simultaneously on a large number of nodes that are decentralized. It means that applications that are embedded in the blockchain are practically unstoppable. On this world computer, there is no admin that has the authority to stop, edit, or sensor applications, as is possible on ordinary computers and servers. This is truly a unique opportunity that has spawned the emergence of a whole series of exciting projects that are believed to change the world.

On the Ethereum blockchain, a special Ether token is installed as a reward for interaction confirmation with a blockchain, miners, and that Ethereum Blockchain users pay for the use of this technology.

On the other hand, Ethereum is designed to be flexible and flexible, so that not only transaction notes can be entered on its chain, but any computer code that can be executed. Ethereum is a programmable blockchain, instead of providing users with a predetermined and limited set of possible operations. Ethereum allows users to create their own operations at any level of complexity. Thus, Ethereum serves as a platform for many different types of decentralized blockchain applications, which includes, but is not limited. Many of these projects took the form of altcoins - separate blockchains with their own crypts, but with improvements in relation to the original Bitcoin protocol that enabled new features and features. At the end of 2013, the inventor of Ethereum, Vitalik Buterin, concluded that a blockchain that could be programmed to perform any arbitrarily complicated computational operation could be more successful.

In 2014, the founders of Ethereum began work on a new generation blockchain that had the ambition to implement a general and complete trust of a deprived platform for smart contracts. Ethereum's greatest benefits are a wide range of applications and the most powerful development community. This project is the biggest source of new solutions, projects, and technologies. Many of the services currently being developed in Ethereum-based projects will almost all be used because they will be faster, better and more favorable. Speaking about the possibilities of using Ethereum is the corresponding programming language - the application capabilities are theoretically unlimited. However, this does not mean that the application of Ethereum will come to life in all situations.

Ethereum itself is agnostic considering its features and value. Similar to programming languages, it is for entrepreneurs and developers to decide what it will be used for. However, it is clear that certain types of applications will have more benefits than Ethereum's capabilities. These are applications that automate direct interaction between equal actors or facilitate coordinated group activities in the network. A good example is the applications that coordinate the markets for the exchange of services and goods between users or automate complicated financial contracts. Bitcoin allows individuals to exchange money without involving intermediaries such as banks, agencies or governments. Ethereum could have far-reaching influence. In theory, financial interactions or exchanges of any level of complexity could be automatically and reliably executed using the code executed at Ethereum. In addition to financial applications, the possibilities of using Ethereum are possible in any ecosystem in which trust, security, and durability are important asset registers, voting, management. It could be significantly changed thanks to this platform. Either, like any other accepted crypt, can be traded directly or through specialized stock exchanges. Some of the biggest stock exchanges are Poloniex, Bittrex, and Coinbase, but there are many. When it comes to direct trade in crypts, it works so that with a person who owns a crypt, you agree to buy a certain quantity from it at a certain price. The price is usually viewed at Coin Market Cap and at the moment of the transaction. When you are ready, the seller sends the Ethereum you bought to your address, and you pay him a fiat with money or another crypt. It's important to pay attention to your wallet. This is a tool that manages the unique address on which your crypt is located. When you make a wallet, safely save your seed or private key and never give it to anyone. The public key is your address, which is publicly visible on the Ethers and anyone can send Ethereum to it. Ethereum has defined phases of development and significant improvements and is sure to change, but better, Tomislav says about the future of Ethereum. They will become faster, safer, more complex, and more powerful. It will enable larger and more complex applications, more transactions and interactions. Due to the sudden rise in the value of the Ether, which is caused by more speculation than real needs, Ethereum has a lot of critics and sceptics. In addition, it attracts a lot of haters because it is the most advanced of all crypto projects. However, we advise everyone to inform themselves well, and the best source of information is the official website of Ethereum.org, but also Reddit and GitHub. He adds that in the Solid chain they believe that Ethereum is waiting for a bright future, that it will have many useful and valuable applications on it, and that we will use it one day in some form. Our opinion on this issue is probably not overly relevant, but that's why Enterprise Ethereum Alliance is here. It is a group of over 150 Fortune 500 companies that support the development of Ethereum and intends to use it in its business. Among the members are Microsoft, J.P. Morgan, the Government of India, ING, Intel, and recently joined both MasterCard and Cisco. Ethereum deserves a special place because it represents an important and great innovation. The ideal creator of Ethereum is Vitalik Buterin, who is now a genius. He proposed in his whitepaper a new type of blockchain that should correct the shortcomings of Bitcoin. In the middle of 2015, this open-source platform based on the blockchain is launched, which allows developers to create and run smart contracts and decentralized applications. Ethereum is a distributed public blockchain network. However, Bitcoin, along with most other crypts before the emergence of Ethereum, offered one application of this technology - a peer-to-peer electronic system for the exchange of digital money. In addition, Ethereum also enables the execution of the code of any decentralized application. The innovation that enables this is the

Ethereum Virtual Machine (EVM) that represents Turing's complete software that runs on the Ethereum Network. Ethereum is currently based on the Proof of Work (PoW) method as well as Bitcoin, although it announces a change in the method in the Role Proof (PoS). Miners dig a crypto-switch Ether which has the same function as Bitcoin but is also used to perform transactions within this blockchain system, i.e. for payment of commission and for services at Ethereum. This means that Ether is also used for all decentralized applications that work on platforms and commissions within them. Creating Ethereum required considerable knowledge of mathematics, cryptography, programming, and a large number of resources to create a blockchain system. Thanks to Ethereum, the entire process is greatly facilitated by developers with the help of tools that are offered to create decentralized applications or writing smart contracts. Anyone can run the program, regardless of which programming language it uses. Potentially any centralized service can be decentralized with the help of Ethereum. Decentralized applications that work on Ethereum also use all the advantages it has and blockchain itself, so they are immutable, cryptographically safe, protected from hacking attacks, operate on the basis of consensus and can work continuously (Zero Downtime). However, smart contracts and decentralized applications are being written by people and it is, therefore, possible that errors or bugs occur in the code. The problem is that once found a way to exploit the error, there is no way for the exploitation to stop here, in order to achieve a consensus on changes to the basic code. An interesting fact about Ethereum is that within two years since the benefits of this blockchain have been made clear to various industries and companies.

Chapter 8: What Is Ethereum And Blockchain Technology? Potential Benefits Of Ethereum

Ethereum is a blockchain-based, public, open-source distributed computing platform featuring scripting (smart contract) functionality. In simple words, it is a rising star in the world of cryptocurrency. It becomes the second largest currency in the digital world in more than two years and spurring the rise and booming in the value of hundreds of new competitors to bitcoin. Ethereum was launched on 30th July 2015 and its value has been amplified up to 6,800% since the beginning of 2017. The current value of an ether is £358 ($480).

This open software platform is based on blockchain technology that allows the developer to design and deploy regionalized applications. After bitcoin, ethereum is the 3rd most valuable form of digital currency after bitcoin with highest market value. Just like bitcoin, Ethereum is a public blockchain network. There are some technical differences between both currencies, such as Ethereum and Bitcoin differ considerably in capability and purpose. Bitcoin offers a particular role of blockchain technology and a peer to peer electronic cash system for online bitcoin payment.

Bitcoin is employed to track the ownership of digital currencies, and the Ethereum blockchain proves helpful to run the programming code for the decentralized application. In the Ethereum blockchain, miners work to get Ether, rather than mining for bitcoin. Ether is a crypto token to fuel the network. Other than a tradeable cryptocurrency, crypto token "Ether" is used by application developers to pay for services and transaction fees on the network of Ethereum.

More Than Money

Ethereum is not only a digital currency, but it is more than just money. It is a platform based on blockchain with numerous aspects. It features EVM (Ethereum Virtual Machine), the smart contracts and utilizes its currency known as ether for peer-to-peer contracts.

Ethereum's smart contracts use applications stored in blockchain for contract facilitation and negotiation. As the benefits of these contracts, the blockchain proposes a decentralized method to enforce and verify them. Different decentralized phases make it incredibly complicated for censorship and fraud. Smart contract of Ethereum aims to offer maximum security as compared to old-style contracts and bring down related costs.

Ether powers all smart contract applications. Blackchain of Ethereum is based on cryptocurrency and ether is a crypto-asset in the Ethereum wallet. It lets you use and create smart contracts. This system can be described as a solo shared computer run by the network of users and resources are paid for and parceled out by ether.

Implementation of Smart Contracts with Cryptocurrency

Ethereum enables you to produce digital tokens. These tokes are used for the representation of shares, virtual assets, proof (evidence) of membership and much more. Basically, smart contracts can be attuned to wallets and particular exchanges that employ API (a typical coin). It is possible to copy a code from the website of Ethereum and use tokens for numerous purposes, such as the representation of shares, voting forms and fundraising. You may possess a particular quantity of tokens in the rotation or an inconsistent amount on the basis of predetermined rules.

Kickstarter is not Required with Ethereum
Ethereum allows developers to increase funds for numerous applications. For a new project, you will get a chance to seek inductees from the municipal and set up a contract. A fundraising will be continued until the goals are reached or a particular date. The funds are released again to the contributors if the particular goal is not fulfilled, or go on the project if it is prosperous. Kicking out Kickstarter means

a 3rd party is booked along with its rules and they also charge a particular fee. It may include processing charges, Kickstarter may take almost 10% of the budget of a project.

Free from Conventional Organization Structure with Independent Democratic Organizations

Ethereum proves good to source funding and offers a legislative edifice to get an exceptional idea. You can get proposals from other people who supported your project and hold votes on the possible procedures. It allows you to skip the expenditure of a conventional structure, such as completing paperwork and hiring managers. Ethereum protects a project from external influences, while it's decentralized network make sure to avoid any downtime.

There are numerous small aspects that make both blockchain-based projects different. Average block time of bitcoin is almost 10 minutes, but the Ethereum aims for 12 seconds. This instant time is enabled by the GHOST protocol of

Ethereum. A quick block time indicates that confirmations are quick. There can be more orphaned blocks. Another difference is the monetary supply of both cryptocurrencies. More than 2/3rd of bitcoins are already mined with the majority going to initial miners. Ethereum elevated its promotion capital with a presale and almost ½ of its coins. These will be minced by its 5th year of existence.

The incentive for mining Bitcoin splits almost every 4 years and it is presently valued at 12.5 bitcoins. On the other hand, the rewards of Ethereum miners are based on the proof of its algorithm known as Ethash, with five ether provided for every block. Ethash is a tough algorithm that inspires decentralized mining by folks instead of using more centralized ASICs similar to Bitcoin.

Ethereum and Bitcoin cost their transactions differently. It is known as Gas in Ethereum and the cost of transactions depends on the storage needs, bandwidth usage and complexity. In Bitcoin, the block size is important to determine the limit of transactions and they equally compete with each other.

Ethereum has its own Turing complete interior code that shows that anything may be calculated with sufficient time and computing power. Bitcoin misses this capability, but there are particular advantages to Turing-complete. Its complications may bring security complexity that contributed to DAO outbreak in June.

Bitcoin and Ethereum are two different beasts and have their own aspects. In reality, both have particular differences and intentions. Bitcoin emerges as a steady digital currency, and the Ethereum aims to incorporate more, with ether is just a component of smart contract applications.

Understand Blockchain Technology

Blockchain technology has the capability to optimize the universal infrastructure to deal with universal issues in a particular space. Nowadays, everyone is talking about blockchain because this concept has keyed up a commotion in the financial industry.

The blockchain is considered as a public ledger of all bitcoin transactions that have ever been performed. One block is an important part of blockchain to record some or all current transactions and after completion, these all go into the blockchain as a permanent database. Bitcoin is a ground-breaking payment network and a unique kind of money. It uses peer-to-peer technology that works without financial authorities and central banks to manage transactions and bitcoins are issued collectively by a network. This currency is an open source and it is available to the public. Blockchains technologies are used to control and record the transaction of bitcoin and Ethereum.

Once an existing block completes, a new block will be generated. These blocks are closely linked to each other similar to a chain in a proper linear and chronological order. Each new block contains a dash of the previous block. In order to use traditional banking as an analogy, the blockchain has a history of previous banking transactions. You can find bitcoin transactions in a chronological order, similar to bank transactions. Blocks can be taken as individual financial statements. Blockchain has a complete record of every transaction of bitcoin. It provides insight, facts, the value of patents and other essential points.

Some developers believe in starting looking at the creation of various blockchains because they don't want to depend on a solitary blockchain. Sidechains and parallel blockchains will be helpful for tradeoffs and enhance the scalability with the use of independent and alternative blockchains. It will increase the chances of innovation.

Example of Blockchain

To understand the blockchain concept, there is an example of a product known as Gyft. It is an online platform to sell gift cards where the customers can redeem, buy and sell gift cards. This business is a partnership between a 44-year old merchant or FinTechn organization First Data and the infrastructure of blockchain provides Chain to provide gift cards of SMBs with the help of blockchains. This product will be rolled out and become a solid example of blockchain-based modernism that has no connection with bitcoin. It is considered as a part of blockchain because a majority of SMBs doesn't have any gift card program and POS that are installed at SMBs may not accept them.

It is really expensive to offer a program for the gift card and hard to notice the immediate benefits. It can postpone fulfillment for a retailer, but large retailers can understand it in a better way or accept gift cards. Blockchain enables Gyft to offer a gift card solution for the customers of SMB.

Non-reputable pseudonymous transaction requests

It is difficult to testify the authentication of a party demanding claim on the digital currency bitcoin. There should be a system to know the actual status of the transfer.

The bitcoin blockchain is answering all these questions. Bitcoin blockchain promotes the use of arithmetical signature algorithms to solve some important issues regarding identity, authenticity, and ownership. This information is required to transfer bitcoins. In the absence of DSA, you are in need of a chief authority to manage credentials of every party and authenticate each transaction on the behalf of parties and recipients have to trust the 3rd party about the validation of transaction.

While using "DSA/cryptography" the details should be verified, such as for define hash of payload that needs to be signed, which includes having a well-defined serialization schema for its messages. As well as well-defined representations for signed messages. The information can be accessed in the blockchain block.

A domain model that fits the problem need

If one ledge contains events for state transitions in the domain model, in this case, the bitcoin needs a particular state transition. It is an ability to move a bitcoin from one place to another.

An algorithm to generate and distribute the supply of Bitcoins – eminently attained by gratifying miners who participate in the P2P network by evidence of the work, angling it towards key adopters to bootstrap the network.

You can get the advantage of future proofing to some extent by implanting the deduced scripting linguistic in the business scheme to design the most complex transfer contract.

There are a few things we can observe from these innovations. Firstly, each of these 3 problems that were solved could have been solved independently.

Bitcoin is not the only system that has Byzantine fault tolerance and these systems may not use cutting-edge cryptography or work with cryptocurrency. Many high availability systems use similar consensus algorithms such as PAXOS to achieve a consensus in a distributed ledger.

Careful use of digital signing can be used in isolation, as it does, for example, in the signing of digital software by publishers to achieve the non-repudiation of the software it publishes (i.e. Microsoft really was responsible for that update) and for clients to know it can be accepted (so auto-update can safely run on your PC). A ledger is not required for any domain logic. It's just saying the software you have really is from whom it came from.

As for the domain model, the government can replicate the capability of bitcoin by just keeping all protocols and transfer rules same, but run it on its own private servers. You authenticate through their servers to check balances and make transfers. It is basically Paypal or you may prefer Bitcoin for everything including crypto and Byzantine error tolerance, but change the rules for the distribution of coin, such as give it to yourself to start a ripple.

One final note on the scripting. The effort most famously taking this to the next level is Ethereum where smart contracts are being built on their blockchain. But hopefully, the Ethereum realizes that they are increasing the capability of bitcoin to distribute it via a P2P network, they could equally be running it as a private service to sell to clients.

The point with all this is that there is a choice in all these innovations. The blockchain is not the only solution to solve numerous problems. You need to pick and choose the pieces as you require them for your problem.

And the best example of that is Bitcoin. Bitcoin is clear on its goals to create a decentralized cryptocurrency and the blockchain technology behind it has been fitted closely to fit that goal.

Chapter 9: What Is Ethereum?

You've probably already heard many phrases that have been coined since cryptocurrency was first introduced. While the concept is still relatively new there is much to learn about it and to understand Ethereum, you need to first understand a little about how this new currency evolved.

Before Cryptocurrency

Prior to cryptocurrency, the world functioned on a series of centralized networks. All records of transactions of any type were usually held on a single centralized server that was managed by an institution of some sort.

Our financial institutions are perfect examples of these new structures. You went to your bank for example, and you made a deposit. The record of that deposit was kept on a single ledger they maintained on a single server. Any transactions made were automatically uploaded to the server via the Internet.

Over the course of many years, these servers were set up for any number of things. The banking industry was a primary user of those servers, but eventually, this system evolved to accept other forms of data that went beyond monetary transactions. One server may hold data on reviews for different businesses, another may hold a series of images uploaded by the users, and another could be responsible for managing transactions related to property sales.

The key point is that whatever transaction was made, the data was held on a server that was owned and operated by a middleman who had his or her own interests to look out for. Therefore, any data held on the server remained under the control of the third party, not yours.

The "server," is actually just a really upscale computer setup with a host of backup systems and lots of memory for massive amounts of storage. Whenever you needed to access your funds to pay a bill or to make a purchase, the merchant you're dealing with would put your information (credit or debit card) into their computer system and then it would connect and talk with the server. If you wanted to make a purchase, it would verify that you had enough funds to cover the purchase or if you needed to pay a bill, it would make sure that the ability to make the transaction was possible. It would then go through the steps to transfer those funds from your account to the account of the client.

While this system has worked for many years, there have been some significant problems with using the centralized server system. First, whoever controls the server controls your assets. They have an incredible amount of power that goes beyond what a third party participant should have. A bank can without warning freeze your assets, lend your money to someone else, or close your account. All of these actions can be done without the need to notify you first on their decision.

Another issue that presents itself is the lack of privacy. Not that your bank is going to advertise your activity to the world but by following your transaction history, they can locate where you are, what you are purchasing, and even the kind of things you're involved in. These institutions often collect this information and sell it to others without your knowledge.

The Blockchain Technology

For years, we had managed to function pretty well with this type of digital currency. However, when the financial crisis of 2008 hit, many began to try to devise ways to eliminate the middleman altogether. They came up with what we now know as Bitcoin, which allows peer-to-peer transactions.

The process is simple. Rather than all the transaction data maintained on a centralized server, it is copied and then distributed to thousands of computers maintained throughout the network. So, rather than one copy of your activities, there are thousands of copies spread out all around the globe. Now, when a transaction is made, the data is uploaded to this network of thousands of computers, and for the transaction to be complete, these systems must all be in agreement. Once completed, each computer in the system will add the transaction to its history in what is called a "block," which is why you often hear the term Blockchain Technology.

Whether you're dealing with Bitcoin, Ethereum, or any other type of cryptocurrency you are working with Blockchain technology. The Blockchain is at the very core of this new method of exchange. Undoubtedly, one-day history will show this to be one of the greatest innovations of the 21st century, and because of it, the blockchain has become a major catalyst for change.

This system is nearly impossible to break because there is no single point of potential failure. The transactions are all made based on an agreed upon system of rules already established. To change those rules, one would have to get every single one of the thousands of computers (nodes) in its network to agree on those changes making it practically impossible to do.

So, what exactly is the blockchain? It is the platform that allows digital information to be distributed to a wide range of users. This may not seem like a really big deal, but this new means of transferring information has become the backbone of the cryptocurrency world.

Blockchain technology was originally created with Bitcoin in mind, but once introduced, many found it could be easily adapted to fit with other digital currencies as well. While you don't need to know all the details of how it works, you do need to understand its purpose and primary functions.

It is important to understand that digital currency existed long before Blockchain Technology, but it is the blockchain that gives the users an alternative means of transferring value without the aid of a third party.

With the blockchain, regardless of the currency, you are trading, all transactions are recorded in a public ledger, which makes it possible to authenticate each one and prevent fraud no matter where the data is sent. Because of its public ledger, it is entirely transparent.

So, What is Ethereum?

Ethereum uses the same Blockchain technology but takes it a step further. With Bitcoin, for example, when you want to transfer funds, the amount of data you need to communicate is very limited. It includes the account you want to send the funds from, the account the funds are going to, and the amount. In most financial transactions this is enough, but Ethereum wanted to have a broader reach so it could be used on a much wider scale.

In addition to the data needed to effect a transaction, Ethereum adds on an additional field called a "bytecode," which allows the user to literally add more code to the transaction. Basically, the user can create a set of rules that can dictate how a transaction can be done. These codes can literally tailor any transaction to meet the demands of the parties involved in the transaction.

These bytecodes create what is called "Smart Contracts" and can be used to create all sorts of complex codes and applications that can influence the how, when, where, and why of every transaction made on the Blockchain.

This opens the door to a wide range of possibilities that can go far beyond just making a monetary exchange.

Ethereum's Virtual Machine

This is all made possible by using the Ethereum Virtual Machine (EVM), a software program that enables anyone to run the program no matter what programming language they use. Because of EVM, Ethereum has the potential of being used in thousands of different ways on a single blockchain.

How Can it be Used?

In light of its easy adaptability, Ethereum can be used with any service that utilizes a decentralized platform. It can be used for obvious services like bank loans and purchases, but it can also be used to manage voting systems, property transfers, and more.

This fact alone makes Ethereum appealing to all sorts of people. The benefits are not to be underestimated. First, without a centralized platform, no third party can interject changes to any agreement, including the government. It is tamper-proof as all the applications are based on a network built around the concept of consensus, it can't be censored and is one of the securest forms of transactions in existence today. With no single point where failure can compromise the system, the application is well protected against fraud.

Ether

Ether is the cryptocoin that runs the Ethereum blockchain. It works like fuel in your car. Without fuel, the car will not operate. Ether is the way you pay for the transactions made on the Ethereum network. If you choose to mine Ethereum, it is Ether you will be working for.

Chapter 10: How Ethereum Works

In this chapter, you are going to learn how the Ethereum network works and how concepts like smart contracts, DAOs and DApps (decentralized applications) function on the network.

Just like Bitcoin, the Ethereum network is based on a shared and immutable public ledger. This ledger is distributed among all the computers (nodes) that form part of the network. Each node holds the most recent version of the blockchain, constantly checking in with other nodes to ensure that it updates its copy to reflect any changes made on the blockchain.

However, there is a major difference between the two blockchains. The Bitcoin blockchain simply keeps a record of all Bitcoin transactions. Since Ethereum is like a huge distributed super computer on which other applications can be run, it must not only keep a record of all Ether transactions, it also needs to keep track of the latest states of all the applications running on it and the smart contracts built on it, their addresses and the unique conditions that determine the execution of the smart contract.

Smart Contracts

I have mentioned the term smart contract severely so far, and you might be wondering what they are. Smart contracts are a new form of transaction protocol that was made possible by the invention of blockchain technology. In their simplest form, smart contracts are basically contracts that can execute themselves without the need for human intervention. They are computer programs that set out the terms of a contract and that automatically and autonomously execute these terms once certain pre-defined conditions are met. Instead of being enforceable by law as traditional contracts, smart contracts are enforced by computer code. Below are some properties of smart contracts:

Smart contracts are built and executed on the blockchain.

The execution of smart contracts is determined and verified by computer code.

The execution of smart contracts leads to a change in the state of the blockchain.

Where Did They Originate?

Though smart contracts have only started being implemented just recently, they are not an entirely new phenomenon. The idea of a smart contract was first conceptualized in the early 1990's by Nick Szabo, a cryptographer and programmer who also made great contributions to the development of cryptocurrencies. Despite having conceptualized the idea, Nick Szabo was unable to implement his idea because smart contracts had to be deployed on a blockchain, which had not been invented at the time. The implementation of smart contracts became possible with the invention of blockchain technology by Satoshi Nakamoto, Bitcoin's mysterious inventor. Bitcoin was the first cryptocurrency to use smart contracts. However, the smart contracts used in Bitcoin are very limited. They can only handle conditions that are related to payment transactions. The development of Ethereum expanded the capabilities of smart contracts, making it possible for them to be customized and deployed on all sorts of transactions.

Are They Really Necessary?

Before looking at the importance of smart contracts, we should first consider the role played by traditional contracts. Contracts emerged as a way for people who do not trust each other to transact with each other. They are a way for the parties involved in the transaction to protect themselves from being cheated or defrauded. For instance, before you do some work for a client, it is the norm for both of you to sign a contract showing the agreement between the two of you. This way, if one of you does not fulfill his part, the contract acts as proof of your agreement, which can be enforced by law.

On this front, smart contracts are not any different from traditional contracts. Their role is to allow transactions to be carried out between two parties in the absence of trust. However, the enforcement of smart contracts is a bit different from that of traditional contracts. For instance, let's look at our previous example where you signed a traditional agreement before commencing work for a client. Sometimes, the client might fail to pay your dues after you have provided your services. In this instance, your only recourse would be to go to a court of law, which would then order the client to hold their end of the deal or risk being fined or sent to jail. This means time wasted and extra expenses for the legal fees. With smart contracts, there is no risk of non-compliance. Once the predetermined conditions are met (provision of services), the smart contract automatically and autonomously executes the terms (payment for service rendered) as stipulated within the contract.

How Do They Work?

Like most computer programs, smart contracts use an 'IF/THEN' logic. If a certain condition is met, then a certain action takes place. For instance, in the above example, the terms of the contract would be IF service is provided, THEN payment gets released to service provider. Remember, I mentioned that smart contracts can be customized for any kind of transaction. This means that you can include as many IF/THEN conditions as necessary as part of a smart contract.

For a smart contract to be set up, there are a number of things that have to be done. The first one is the definition and submission of the subject of the contract. The smart contract needs access to whatever is being transacted under the contract. In our example, the subject of the contract would be the payment being made by your client. This allows the smart contract to lock the payment (subject) until you meet the conditions of the contract, at which point it will automatically release the payment to you. Secondly, there is need to define the terms of the contract. These are the conditions that will trigger the automatic execution of the smart contract. In our case, the terms would be the service your client has hired you to provide. Once the terms are defined, both parties must sign the contract with their digital signatures. Just like traditional contracts, a smart contract does not become valid until it is signed by all the involved parties. However, unlike the traditional contract which you sign with your pen or official stamp/seal, a smart contract is signed using your private key, which controls the access to your digital wallet.

Advantages Of Smart Contracts

Smart contracts have several advantages over traditional contracts, which include:

Autonomy – One of the biggest advantages of smart contracts is that they are self-executing. This eliminates the need for third parties, such as lawyers and courts, and the fees and time resources needed for intervention by these third parties. Smart contracts also eliminate the risk of the third party being corrupted to change the terms of the contract.

Trust – This is related to the previous point. Since the smart contract is autonomous, you can take part in non-trust-based transactions with the confidence that all parties will hold their end of the deal.

Safety – If you were to lose your copy of a traditional contract, then you have no way of proving that you actually entered into a contract with the other party. Smart contracts, on the other hand, are stored on the blockchain, which is permanent and immutable. This means that your contract is as safe as can be.

DAOs And Decentralized Apps (DApps)

Apart from smart contracts, there are two other new technologies that are closely tied to the Ethereum network. These are DAO's and DApps. These two concepts can be a bit confusing for beginners. This is because, on the surface, they appear to be similar. However, they are not one and the same thing. In this section, let us take a look at the differences between DAO's and DApps.

Decentralized Autonomous Organizations (DAO's)

Before we get to the actual definition of DAO's, let us look at a thought experiment that was introduced in 2009 by Mike

Hearn, a former Bitcoin contributor. In the thought experiment, picture a self-driving cab that drives around the city looking for passengers. At the end of every day, it uses the profits earned from the day's work to refuel itself and to automatically pay for its insurance. Every Saturday morning, it drives itself to the mechanic for servicing. All this happens without human intervention. Does this sound unbelievable? This is the kind of revolution set to be brought about by DAO's.

Also known as Decentralized Autonomous Corporations (DAC's), these are organizations that run autonomously by following conditions prescribed by smart contracts. All the rules governing the operation of the organization are encoded into smart contracts and put on the blockchain. This eliminates the need for hierarchical management to enforce the rules of the organization. In other words, a DAO is similar to a normal company, only that the DAO operates on the blockchain environment, with the organization's rules being enforced by smart contracts. The idea of DAO's came about after people in the crypto space realized that blockchain technology could do away with managerial wastage the same way Bitcoin made it possible to do away with middlemen in the financial sector.

It is important to note that there is a difference between DAO and The DAO. DAO represents the type of blockchain-based leaderless organizations, while The DAO (the one that got hacked) was the name of one of the first organizations of this kind.

So, if there is no management in DAO's, how are decisions made within the organization? Since the DAO is decentralized, the power to make decisions is devolved to all members of the organization. Any member of the organization is allowed to make a proposal. The rest of the members in the organization then vote in favor of or against the proposal and the decision is made based on consensus. In the normal world, such a decision-making structure would be tiring and time consuming.

However, since this happens on the blockchain, with the members of the organization being represented by computers, decisions can be made much faster.

DAO's make it possible for people from all over the world to work together towards a common objective without the need for third parties to instill trust. While this is nothing new, DAO's offer a faster and more efficient way of doing it, similar to how smart contracts are more efficient than traditional contracts.

Anyone can invest in or take part in a DAO. All you need is to have a wallet and be well versed with the process of purchasing Bitcoins or Ether (this will be discussed in greater detail in Chapter Four). To invest in a DAO, you simply need to purchase the tokens related to that DAO. The tokens act as shares, giving you a right to vote on the affairs of the DAO or a claim of the profits made by the DAO. Similar to owning shares, the more the tokens you own, the more voting power you have.

Advantages Of DAO's

Since the aim of DAO's is to make the running and operation of ordinary organizations more efficient, DAO's have a number of advantages. These include:

Decision making is based on consensus and all the members of the organization have the opportunity to make their contribution towards the running of the organization.

The lack of hierarchical structure means that any member can put forward their proposal for consideration by the entire organization.

The rules governing the operation of the DAO are pre-written and cannot be changed, which eliminates the risk of the goals and objectives of the organization being corrupted.

DAO's are transparent, since all the rules and financial transactions are permanently and immutably recorded on the blockchain, which is accessible to all members of the organization.

In order for a member to put forward a proposal, they need to spend some tokens. This incentivizes the members to think through their decisions and keeps the organization from being spammed with ineffective solutions.

Disadvantages of DAO's

Since they are a relatively new concept, DAOs have not been fully tried and tested.

Since DAO's are essentially based on code, any errors or bugs within the code can be exploited by people with malicious intentions, such as was the case with The DAO hack.

DAO's devolve decision making power to the masses. Some experts feel that this is not a very wise move, especially in financial matters.

In order for DAO's to gain mainstream adoption in the real world, there is need for legal frameworks to govern their operation, something that is non-existent at the moment.

Decentralized Applications (DApps)

Decentralized applications are still a fairly recent concept, so there is no unanimous definition that can be used to describe them. They are a new breed of apps that cannot be stopped, that have no downtime and that are not owned by any one person. Despite the lack of a unanimously accepted definition of 'decentralized applications', there are certain properties that any application must have before it is categorized as a DApp. For an application to be considered as a DApp, it needs to be open source, it has to be built on top of a public and decentralized blockchain, it needs to use tokens that are cryptographically generated, and it needs to have an inbuilt protocol for ensuring consensus.

Sometimes, decentralized apps are confused with smart contracts. To understand the difference between DApps and smart contracts, we can compare DApps to websites that are blockchain enabled, while smart contracts are the technology that allow these websites to access the blockchain. In a normal, traditional website, the front end of the website (what you see) uses API's to connect to the database containing the files that make up the website. With DApps (blockchain enabled website), the front end of the website uses a smart contract to connect to a blockchain. This means that smart contracts are a part of DApps.

Another major difference between DApps and traditional applications is that traditional applications are hosted on centralized servers. This means that if the server fails, then a traditional application would experience some downtime. DApps, on the other hand, are hosted on decentralized peer to peer networks. This means that it is impossible for the DApp to experience downtime, since it is practically impossible for the entire network to fail simultaneously.

Chapter Summary

In this chapter, you have learned:

Smart contracts are basically contracts that can execute themselves without the need for human intervention. They automatically and autonomously execute these terms once certain pre-defined conditions are met.

Smart contracts were first conceptualized in the early 1990's by Nick Szabo.

Bitcoin was the first cryptocurrency to use smart contracts, though the smart contracts used in Bitcoin are very limited.

The development of Ethereum expanded the capabilities of smart contracts, making it possible for them to be customized and deployed on all sorts of transactions.

Smart contracts are similar to traditional contracts, only that they are executed on and enforced by the blockchain. This makes them more effective and efficient.

Smart contracts use an 'IF/THEN' logic. If a certain condition is met, then a certain action takes place.

For smart contracts to work, they require a subject, the terms of the contract, and the digital signatures of the parties involved.

Smart contracts provide autonomy, trust and safety.

Decentralized Autonomous Organizations (DAO's) are organizations that run autonomously by following conditions prescribed by smart contracts.

A DAO is similar to a normal company, only that the DAO operates on the blockchain environment, with the organization's rules being enforced by smart contracts.

Since the DAO is decentralized, the power to make decisions is devolved to all members of the organization.

Decentralized applications (DApps) are a new breed of apps that cannot be stopped, that have no downtime and that are not owned by anyone.

For an application to be considered as a DApp, it needs to be open source, it has to be built on top of a public and decentralized blockchain, it needs to use tokens that are cryptographically generated, and it needs to have an inbuilt protocol for ensuring consensus.

Since DApps are run on decentralized peer to peer networks, they cannot suffer downtimes resulting from server failures.

In the next chapter, you are going to learn about gas and its role within the Ethereum network.

Chapter 11: What Is Cryptocurrency?

We're going to jump right into this book by assuming that you've got little to no understanding of cryptocurrency or what it is. Because of this, we're going to jump right from the beginning and explain some of the bigger concepts underlying cryptocurrency. If you're already familiar with what cryptocurrency is and its foundations, then feel free to go ahead and move forward to chapter two, which is where we'll start discussing Ethereum in detail.

So what is a cryptocurrency and what is its purpose? This question can be a bit of a tricky one to answer. The truth is that cryptocurrencies fill a very particular niche, and they can be an astounding one.

So here's the thing: to understand why cryptocurrencies need to exist, you need to consider for a second how you exchange money right now. When you pay for things in real life, what channels do you normally use? For the most part, people will use either debit or credit cards, or cash. There's a big difference between these two different modes of exchanging money, though.

When you transfer money by using a debit or a credit card, you aren't physically giving somebody money. What is happening is that you are storing your money (or the promise of money, in the case of credit) somewhere, and that can then this is then routed through the bank towards the person who is to receive the funds - generally to their own bank. This means that this method of exchange is ultimately based upon the usage of a bank as an intermediate.

However, in life, you have another choice. You can exchange cash with somebody. The cash is a physical manifestation of currency. This means that only one person can hold it at a time and it can go from one person to another without somebody verifying the transaction taking place. In other words, if Bob were to give five dollars to Carl, then Carl were to give that same five-dollar bill to Dave, then Dave was to give that same five dollars to

Edward, then there is no way to trace the lineage of the bill. That is to say that it could only be certain that Bob held the bill first (the bill's point of origin) and that Edward held it last (the bill's final destination). There is no way of keeping up with things from one place to another.

In internet-based transactions, though, everything must take place through a third-party. Or, it had to at one point, anyway. If you wanted to exchange currency online, it would have to go from point A to point B through the use of a third-party, such as a bank or some other form of online currency exchange such as PayPal.

The problem with this is that agency over the transaction is, by the nature of this transaction, delegated to the third party. This is obviously unfavorable for many people. A lot of people think that the idea of their money being monitored by a third-party is very untenable or undesirable in general.

The fact that online transactions had to be carried out by a third-party is known as the problem of trust because it requires both users to trust that the third party is going to carry out the transaction as described with no funny business.

However, this sets the stage for a natural progression: moving past the problem of trust.

The problem was that there was no way to directly transfer cash from one computer to another. This is because there was no physical asset to represent the money. Take, for instance, the dollar bill; a dollar bill is a tangible physical asset that one person can hand to another person. At that point, the dollar bill is now physically in the hands of somebody else.

This sort of direct transfer of funds without any sort of third-party has long been mulled over by the technological community. For the most part, it was long considered to be impossible. This was as a result of the double spending problem.

To understand the double spending problem, you have to understand how computers and files work. The most immediate thought of direct cash transfer over the internet would be to simply move it from one computer to another. However, this doesn't exactly work.

Files may never be 100% unique, and even a file with unique properties may be replicated with those same exact qualities. This means that to a certain extent, a file may never truly be passed on. Unlike cash, where there is only one tangible instance of a dollar bill, this one tangible instance doesn't exist for files on a computer. This means, due to the intangibility, that the same digital bill could be duplicated and used more than once without very much trouble. This conundrum is often referred to as the double spending problem.

For a long time, the double spending problem was mulled upon. Solutions were few and far between. This was, of course, until a guy (or group of people) going by the alias Satoshi Nakamoto published a whitepaper called Bitcoin: A Peer-to-Peer Electronic System. This was the beginning of the wave which would carry cryptocurrencies forward.

This paper posited the usage of peer-to-peer technology to drive a decentralized currency. The notion of peer-to-peer means that the technology is divvied up amongst multiple different sources with no one absolute origin. This means that the currency would ultimately not be centrally controlled but rather delegated amongst the people who are using it.

There are multiple reasons that this is considered favorable. We'll go into each in a little detail.

One reason is that having a peer-to-peer currency is seen as favorable is that, much like cash, it would be very hard to trace. Without a centralized authority holding dominion over the currency and the transactions of that currency, it is extremely difficult for anybody to control what happens with that currency. This means that, in a sense, a digital peer-to-peer currency could be used in much the same traceable way as a fiat currency.

Perhaps the most intriguing part of the development of cryptocurrency and Bitcoin, in particular, was the implementation. The exact implementation of cryptocurrency typically involves the development of something called a blockchain. Blockchain was a revolutionary new idea, and it's important that we talk about all of this before moving further because to understand Ethereum, you have to understand a blockchain and how it works. Essentially, think of it this way: when you go to a bank, they keep a record of every single transaction. All of these transactions are kept in a series of ledgers. They'll contain things such as amount withdrawn and the time of withdrawal, as well as any sort of general account transactions. The important thing to note here is that these things are marked down by the bank.

In effect, the blockchain isn't that much different from an ordinary ledger. However, the key things which make it really neat are firstly, the unique digital implementation which opens them up to be used for many different purposes and secondly, the fact that they are entirely decentralized (usually) and are therefore not under the dominion of any one person.

So how does the blockchain work? Well, it's not exactly one big ledger; it's more like a bunch of little tiny ledgers. Note that these ledgers don't necessarily have to contain transaction data per se; rather, a ledger is just a manner of which any sort of data can be recorded, usually including some sort of date-time information but not necessarily.

Each of these tiny little ledgers is called a block. When transactions take place, they are written to a new block. This block will grow until it is solved. So, what do I mean by "solved"?

Solving is the cryptographic process by which a block can be verified as holding real and true information. Solving is generally a multi-step process, comprised first of verification and then of hashing.

The verification process occurs first. Most cryptocurrencies, like Bitcoin and Ethereum, are driven by people called miners. These miners will have the special software set up that allows them to serve as a node for their given cryptocurrency. What happens is that all nodes will have a copy of the current blockchain in its entirety. The nodes will check against one another to determine what version of the current block is the right one. For the most part, this is a straightforward process; due to the nature of the peer-to-peer network, new transactions are automatically detected and written by the nodes. This means that the blockchain is pretty much impenetrable because, in order for one to skew the reality of the current block, they would have to have over half of the current processing power of all nodes. For larger cryptocurrencies like Bitcoin and Ethereum, this is simply unthinkable.

Now, all computers which have the cryptocurrency installed and use it are constantly running and utilizing the blockchain as well. It's just that not all computers are actively dedicating themselves to mining. Some would still refer to these computers as nodes because they're still doing part of the verification work for any given block by offering up their own consensus on the current block.

After the verification process comes to the hashing process. This is the process by which a given block becomes a part of the blockchain. Basically, imagine that all the blocks are linked together by something like chain-links. To link a new block to the chain, you have to find the place where the chain-links may intercept one another. This is done in many different ways depending upon the specific currency.

In Bitcoin, for example, the hashing is extremely math-heavy and is called proof-of-work. This means that the block is verified by nature of the fact that enough nodes have done enough work to verify that the block is ready to be used. This is done by a system of complex mathematical equations that are performed to find a key which links the current block to the former. This is far easier said than done and, in the example of Bitcoin, can include up to trillions of transactions to find one key.

Ethereum also uses a proof-of-work algorithm to verify its blocks. However, its proof-of-work algorithm is slightly different. The problem with many cryptocurrencies, such as Bitcoin, is that their proof-of-work algorithm can be manipulated in such a way that normal consumer computers don't have a chance of keeping up if they want to mine blocks. This means that to mine bigger cryptocurrencies, like Bitcoin, you'll most likely need to buy a specialized rig to have any hope of making a profit. Indeed, Bitcoin itself has a built-in difficulty scaling to keep blocks from being verified too fast. Ethereum does as well, but the nature of Ethereum's algorithm means that you most likely aren't going to need specialized hardware to mine it.

In fact, Ethereum's algorithm was specifically designed to combat the issue of mining centralization. What this is is where a small number of people can procure enough hardware and enough of a mining setup that they start to have a large share of the mining power and thereby are the ones raking up the profit and also have the potential to manipulate the network. While Ethereum's proof-of-work algorithm is far from perfect, what it does mean is that you can still easily mine it while using consumer hardware - that is, hardware which isn't necessarily specialized for the purpose of Ethereum mining. It does so by being memory hard - this means that the algorithm is based around moving data around within your computer's memory banks instead of its ability to perform raw calculations. Ethereum's mining algorithm was specially designed just for it.

A little bit later in the book, we'll be covering how to mine, buy, and keep Ethereum's cryptocurrency, Ether. However, for right now, it is simply imperative that we cover the basics of what Ethereum is and what makes it different from other cryptocurrencies.

Chapter 12: Money, Cryptocurrency And Ethereum

Money

Everybody has at least a vague idea of what money is. Most people immediately think of coins and notes. All sorts of things including beans, salt, checks, and gold have been used as money. Money has two features, which are first, it is a store of value and, secondly, it is a means of exchange. In the past, it has been issued by a central authority usually a bank or a government.

Cryptocurrencies

Cryptocurrencies such as Bitcoin, Ether and Ripple use a new idea that money can be manufactured with the encryption techniques used in advanced computer programming. With these techniques is possible to transfer funds and verify that this has occurred. True cryptocurrencies are decentralized by being spread across networks of computers and do not have a central point of issue such as banks and governments.

Although some central banks are investigating this technology, what they come up with cannot be cryptocurrency, as it will still remain centralized at the bank. It is the lack of control by a central authority, such as a bank, that is one of the most appealing features to the advocates of cryptocurrency.

A Short History of Cryptocurrencies

There has been much publicity about cryptocurrencies, such as Bitcoin and Ethereum. Only now is the general public accepting that cryptocurrency is, in fact, a genuine form of money. As a result, the value of some of these cryptocurrencies in dollars has increased enormously.

Bitcoin, the first cryptocurrency, originated in 2008, but for a period of years was seen by the general public as a fringe form of counterfeit used by criminals on the deep web to purchase guns, drugs and other illicit goods. Despite this, the underlying technology of Bitcoin, the Blockchain, was realized as a technology, which had great potential in many activities beyond creating new types of currency. The use of this technology has exploded and as a consequence, the total capitalization of the cryptocurrency market is in excess of $100 billion!

Some Interesting Facts about Cryptocurrencies

In June-July 2017 there were no fewer than 900 cryptocurrencies. The majority of people know about Bitcoin because ransomware attacks in 2017 and earlier, have required payment in Bitcoin. Criminals wanted this so that any payment by a victim could not be traced. It is very difficult, if not impossible, to track transfers of cryptocurrencies.

There are very good websites, such as coinmarketcap, for information about this burgeoning market.

If such a website is visited then Bitcoin will be seen to be the most valuable cryptocurrency with one Bitcoin, in late July 2017, worth more than $2550 and a total market capitalization of Bitcoin of about $43 billion!

The second most valuable cryptocurrency is Ether of Ethereum, with a value of about $200 per coin and a total market capitalization of more than $8.5 billion.

Not all cryptocurrencies have valuable coins, to prove this we must have a look at other cryptocurrencies. A good example is Ripple, which is the third most valuable cryptocurrency; the market capitalization of Ripple is about $7 billion, however, the value of one Ripple coin (XRP) is only $0.20.

Some cryptocurrencies have very small capitalizations. One such is XenixCoin cryptocurrency with a capitalization of less than $300! We will explore the quality and worth of cryptocurrencies later.

Cryptocurrency Compared with Fiat Currencies and Stocks

Currencies used in everyday life such as dollars, euros, yen, and renminbi are named 'fiat' by the people wishing to contrast them with cryptocurrency.

Despite the word currency in cryptocurrency, there is a greater resemblance between stocks and cryptocurrency than there is between cryptocurrency and fiat money. If a crypto currency is purchased then the acquisition of the coin is also the purchase of a technology stock and an entry in a digital ledger called a Blockchain. Blockchain will be shown to be all-important when we consider Ethereum.

What Is Ethereum?

First and foremost, Ethereum is a Blockchain technology. It has a cryptocurrency associated with it called Ether. Most people when they read or hear Ethereum mentioned see it as nothing more than a cryptocurrency, but it is much more. In the next chapter, we will examine what Blockchain technology is.

There has to be trust that these companies will protect your data. Generally, they do and successfully. They employ huge teams of very clever people to enable this. Unfortunately, hackers are ever increasing in number and sophistication and there have been numerous cases where valuable data such as credit card details, unreleased films and compromising photos of celebrities have been stolen.

With Blockchain technology, the need for this central control is removed and safety vastly improved. Some argue that the Internet was always meant to be decentralized and Blockchain technology is one of the ways by which this is being attained.

Ethereum can be seen as a platform enabling this.

Ethereum's stated aim is to use Blockchain technology to do away with all third parties on the Internet, the people and organizations that clip the ticket. Third parties involved in the storage of data, real estate, the movement of finance and many other transactions.

In order to carry this out, ethereum has thousands of what it calls nodes spread throughout the world, replacing the servers called the cloud.

Is Ethereum Money and Is It Different to Bitcoin?

The answer to the first of these is no and to the second yes. Ether is the cryptocurrency that is associated with Ethereum. As a result, Ether is a means of exchange that uses cryptography so that transactions with it are secure and there is control over the manufacture of further units of the currency. Ether is a type of what is called alternative currencies.

Bitcoin is essentially money that uses Blockchain technology. The Blockchain associated with Bitcoin is mainly designed to serve this monetary purpose while Ethereum is a whole platform of Blockchain technology with its cryptocurrency, Ether, only a fraction of the whole platform. Ether to Ethereum can be compared with Excel to Office.

Has Ethereum Any Advantages Over Fiat Money?

As this question relates to money we will restrict ourselves to considering Ether, the "money" of ethereum. Ether has advantages that were recognized when the idea of cryptocurrencies was conceived.

Ether is decentralized and is out of the control of nations or companies.

With Ether spread over a huge network, there is no possibility of failure of the currency as a result of failure at one or even many points. Even if there were these failures Ether would continue.

By using Ether there is much greater privacy than if fiat currencies were used.

By design and concept ether is very easy to use, often a lot easier than using fiat currencies which are burdened with fees for intermediate transactions and greatly slowed by bureaucracy.

It is much quicker to transfer Ether than fiat money.

Because Ether is created with cryptography, the possibility of fraud is greatly reduced in comparison to fiat where credit card numbers are often stolen by hackers and sold on the Internet

Chapter 13: The Rules Of Serious Investing

When you are ready to get started in investing, whether you want to invest with cryptocurrency or some other option, it is important to take things seriously. It is tempting at times to go out there and just start investing and think that this isn't serious. Since most of the investing are going to happen on a computer and the idea of Bitcoin and Ethereum are so new, it is easy to think that you are playing with pretend money. But just like with other investments, you are dealing with real money that you can win or lose when working with Bitcoin and Ethereum and you need to take it seriously.

If you are ready to become serious about investing in cryptocurrency, it is important to understand where to get started. Here we are going to take a look at some of the best rules to remember when you are ready to get started with cryptocurrency investing.

Markets return to the mean

As you are looking at the market for Bitcoin or Ethereum, you will notice that there are some big highs and some big lows and many times when it is just in the middle. But no matter where the market has been at one point, it is always going to go back to the middle again. This can be really helpful to you when determining which investment to make.

For this one, you would need to take a look at some of the charts that are available for the investment you are interested in. See where it is right now and then look to see where the mean is. If the investment is really high right now and is reaching about as high as it usually does, this means that it is going to go back to the middle pretty soon. This may not be the best option for you to pick right now because the investment will go down.

On the other hand, if you notice that the investment is lower than the average and has been there for a bit of time, this may be a good time to make a purchase. It is likely that the investment will go back up soon, and you will make a good amount of money.

Excesses are never going to be permanent

It is sometimes hard for new investors to realize that their profits are limitless on one investment. Yes, you can keep going with investing and if you can move things around and make a lot of money, but each investment option is going to have some kind of limit to how much you can take in. While there are going to be sometimes when an investment, such as with Bitcoin or Ethereum, will skyrocket and make a lot of money in the process. But these trends are not going to be permanent. Instead, they will eventually return back to the mean. A smart investor will realize this and make some plans for when the reversal happens; a beginner will assume that these excesses will continue on forever and they will be burned.

The public will buy the least at the bottom and the most at the top

When you look at the average investor, you will find that there are some trends in the way that they make their investing decisions. These investors are often impressionable, innocent, and don't have the right counsel to help them make smart decisions. They may read the newspaper and pay attention to what they hear on TV, but they don't really know much about the market on their own.

The issue with this option is that by the time reporters tell the news about the market, the move has already occurred and a reversion has started, meaning that they have gotten some outdated information that can ruin their chances. This is why you as a good investor need to learn how to become a contrarian. Being able to do your own research and how to have independent thinking will serve you well. This helps you to make smart decisions that will get you ahead, rather than following the advice from outdated and uninformed sources.

Avoid greed and fear

What this one basically means is that you need to use your head, rather than your emotions, when making investing decisions. It doesn't matter if you are using cryptocurrencies or other forms of investing, the second that you let those emotions get the better of you is the moment that you will lose out on any profit that you hope for. These basic human emotions are going to be the biggest enemy that you try to be successful in investing.

If you want to see success with any type of investing that you choose, you need to learn how to make level-headed and disciplined decisions You have to look at your resources, watch the market, and make decisions based on your strategies. Your emotions should never factor in. There are too many times that people will get out of the market when there is just a tiny dip because they are afraid of losing everything, but then a few days later they find out they missed out when the market goes skyrocketing, in addition, there are people who avoid their strategy and keep going with the market because they want to make more profit, and they miss out when the market suddenly goes down.

The market will have these ups and these downs, but if you have a good strategy in place ahead of time, you can avoid some of these issues. You have to know when it is actually time to get out of the market, both when the market is up and when it is down, rather than just running away because you are scared of where the market is going or you become greedy and want to make more. The market will always turn around so having a strategy and an exit plan for every investment will help save you.

When the experts all agree, go the opposite way

There are a few times when the forecasts and the experts all agree that one thing or another will happen. But this is usually a time when you need to be on alert. It usually means that the market for that trend has become saturated and that everyone in the market who will purchase that investment has already made purchases. It means there are no longer any buyers. Once this happens, the market must go lower so that someone new will make a purchase, since there are likely to be many sellers and very few, if any buyers.

It can also go the other way as well, so always be careful when all the sources start to agree. They will often fall behind and if you go by this advice, especially when the forecast and the experts all agree, you may find that you make the wrong decisions.

Watch for the three stages of the bear market

What some new investors don't understand is that all markets are going to have stages and learning these stages will help you to figure out whether it is the right time to get into the market or not. This is true even when you are dealing with cryptocurrency. When you are working with a bear market, there will be a sharp downward turn, a rebound when some new buyers think the market will get better, and then the downtrend is drawn out a bit more.

A bear market doesn't mean that you shouldn't invest. There are still plenty of investments that do well in these markets, you just need to make sure that you pick out the right option. And this is where cryptocurrencies can come into the mix. These have all done well, and while they do have some ups and downs on occasion, they have not had the bad markets that some of the other stock markets have. But even in the stock market, you will find some investments that do well even in a bear market so just do your research and look around for the best option.

Have your own strategy

There are lists and lists of strategies that you can read about online and in books and many of them are great resources to help you come up with your own strategy. But no matter what these books say, you need to create your own strategy. And since both Bitcoin and Ethereum are new forms of investing, you can have a lot of fun coming up with your own strategy to really see success.

You have to come up with a strategy that is going to fit you. If you like to look at charts and go with trends, this is a good option. If you want to take a look at the company and do a fundamental analysis, then go ahead and do this. The most important thing is that you find a strategy that you are comfortable with and that you stick with it. The biggest reason that people fail in investing of any kind is that they just weren't able to stick with their strategy and kept hopping around.

Getting into investing can be hard, no matter what kind of market you are working in. But understanding the topic that you want to go with, such as picking a company you like, backing up a product you feel passionate about, or something else, is so important when you want to see success. Stick with some of the rules that we discussed in this chapter, and you are sure to see results in all of your investment options.

Invest with others

One way to help reduce some of the risks that you are dealing with is to combine your assets and invest with others. Sometimes you may not have the money needed to get started, such as when you want to help a new startup. Most people don't have that much money and they don't want to put in that much risk when starting out. When you combine together with other investors, you can still get in on the investment without having to lose out on everything if it fails.

This works kind of like a mutual fund. The investors can all share the risks, but they also share the profits too. If you don't have enough money to get in on an investment by yourself, this is a good way to start out. If the company does well, you will earn a percentage of the profit based on how much you invested. But if they do poorly, you will lose some money, but that is shared as well. It is a safe way to build up your portfolio and to start earning money in cryptocurrency.

Chapter 14: What Is Ethereum?

More than anything else Ethereum is a platform for developing blockchain applications. This fact about Ethereum accounts for its importance and is central to the incredible success of Ethereum.

Ethereum's early history: Vitalik Buterin proposed Ethereum in a whitepaper written in 2013. Buterin is an extraordinarily gifted young Russian, born in 1994, who is only 23 years old. In 2000, his parents emigrated from Russia to Canada for better opportunities and where Vitalik received most of his education.

While at school, it soon became apparent that he had a flair for mathematics and computing. His father, who was a computer scientist, taught Vitalik about Bitcoin. Vitalik was fascinated and saw possibilities for blockchain, far beyond its commercial use in Bitcoin, the extraordinarily successful cryptocurrency.

He wrote his paper and ended up leaving his university studies at the University of

Waterloo in Ontario to work full-time on Ethereum. A core team of four members, including Buterin, started the development of the Ethereum software project. Development money was provided by crowdfunding, using payments in Bitcoins.

The Ethereum concept was always lauded for its brilliance, but from the very outset, there were serious questions about its scalability and its security. Ethereum was launched in July 2015 and has proceeded through four versions since then, with Olympic version 0, Frontier version 1, Homestead version 2 and the next version 3, which is Metropolis due late September 2017. There are plans for a version 4, called Serenity.

Concerns about security proved well-founded, as in 2016, an organization called DAO, which stands for decentralized autonomous organization, a set of smart contracts (we will discuss these later) based on the Ethereum platform raised $150 million to crowdfund itself. Sadly, a hacker was able to raid the DAO and steal

$50 million (US). This disaster caused a split in the Ethereum community and led to the creation of two types of Ethereum. Ethereum (ETH), which is the subject of this book, and Ethereum Classic (ETC), a cryptocurrency with a market capitalization of nearly $1.5 billion (US).

The split was the result of what is called a hard fork in blockchain technology. A hard fork is where in block chain technology the protocols for mining and transactions are altered so that previously invalid blocks and transactions become valid or vice versa. The effect of a hard fork is that all nodes must now use upgrades to software to use the new protocols.

A hard fork is similar to a schism in religion but on a technical level. At this point, it is worth mentioning what a soft fork is. A soft fork occurs where previously valid blocks or transactions become invalid. The difference between hard forks and soft forks is very technical, and we will not dwell on this difference.

In this book, we have described the mechanics of blockchain, which makes

Ethereum possible. However, the reader may be wondering why Ethereum? Why not just use Bitcoin? Bitcoin has been around for eight years and is gaining ever more support, why not just use it?

The answer is quite simple. Bitcoin is first and foremost money and fulfills this purpose brilliantly. With Ethereum it is important never to forget the words from the beginning of this chapter,' More than anything else, Ethereum is a platform for developing blockchain applications.' Such blockchain applications are called smart contracts.

Smart contracts are in essence digital exchange processes. The blockchain, which gave rise to transactions using cryptocurrency, is capable of all sorts of value exchanges. With this technology, the checking, execution and security of a vast number of these are possible.

The means of doing this is by the use of dApps (distributed applications), written in computing languages, such well-known languages as C, JavaScript of websites and Python and languages specially designed for Ethereum, such as the Solidity language. In computing jargon, the dApps used in developing smart contracts produce what are called high-level programming objects. The Ethereum blockchain processes these. The technology for this will be explored far more thoroughly in a future chapter, in this chapter it is sufficient to be aware of the existence of smart contracts.

Chapter 15: Gain With Ethereum

It is true Ethereum, also its inherent Blockchain technology that has captured the shareholders' interest in Ether. This chapter

explores the investment fundamentals of Ethereum and its applicability to miners to supply you with the big picture that you will

need prior to investing in Ethereum. Before we delve deeper to Ethereum investment, it is critical that we answer the query,

"What's Ether"?

The Idea of Ether Ether--or ETH--would be the unit of account and shop Value about the Blockchain of records in the Ethereum network. It's the equal of bitcoins (BTC) to the Bitcoin network. While using an economic value (in the meaning that it's a rare

commodity), Ether isn't supposed to be utilized as an alternate money such as the Bitcoin.

Instead, it has been designed as a method Reserve that powers the development of those trying to utilize the Ethereum platform to

produce programs that can generate value for the users. If Bitcoin's worth is obtained from the lack of the commodity and the

safety of the network, subsequently Ether has worth as it's required to implement scripts and smart contracts around the Ethereum

ecosystem.

For this reason, Ether is currently Called the "digital oil" while Bitcoin is the "digital gold". While bitcoin triumphed in

proliferating naturally over time through Bitcoin mining, the Ethereum community wanted to find a means to kick-start the mining

process and attract a foundation of Ethereum technology fans who may help the system to develop.

To reach a Vital mass of Ethereum Programmers, Ethereum's team employed Ether as a motivation to bring the project to life. In

July 2014, Ethers became straight available for purchase on Ethereum.org and more than $18M was increased through that effort. The

point of contention which has emerged lately hinges on the legality of the first sale.

But It's vital to note that to date, no Action was taken against any of the individuals or groups that were involved as well as

other Blockchain growth users who have used this method to community building. Nevertheless, the legal technicalities involved

happen to be acknowledged by Ethereum programmers.

Inflation RateThe Ethereum system has a mechanism for Releasing new Ethers to the network over a given time. Something of

importance to investors that are familiar with Bitcoin and other crypto monies is that there is a gap between the approach from

the Ethereum system along with other crypto currency systems. For instance, from the Bitcoin system, the limitation of all the

bitcoins which will ever exist was set in 21M Bitcoins, a caveat that will require a consensus of nodes to change.

Ethereum, on the other hand, has no fixed Limit on how much of its digital tokens will exist later on. Rather, its development

group wants to use its own investment system in a manner that promotes access by introducing 18M Ethers each year through mining.

They argued that over a specified time, this constant rate of inflation would decrease since the general token supply improved.

As a result of inflationary rates, fresh Participants from the ecosystem will have the ability to obtain the new Ether or mine for

the new Ether if they live in the year 2020 or 2120.

Similarly, the economic model of Ethereum is Slightly different from that of Bitcoin as Ethereum does not halve the financial rewards after four years. It follows that you will receive exactly the exact amounts of Ether each year. However, in Bitcoin, the

financial returns are halved after every four years. When it comes to reserve side of the electronic money, there's no limitation

in Ethereum. Actually, the amount of Ether which can be circulated is limitless.

GasIf Ether is worked as a way to Permit access Into Ethereum's world computer and guarantees its functionality, an economic

structure is also required to limit access. To match Ether and better explain the workings of its own token, Ethereum introduced

the concept of "Gas," a throttling approach that controls (in real time) how much Ether every smart contract expenses.

Gas has a steady value That's presently set At 10 "Szabo", with one Ether being made up of 1M Szabo. The more it takes for the

smart contract to run, the more systematic resources it requires, thus needing more fuel to execute the wise contract. Executing
smart contracts based on the Ether limit or the Gas throttle is a market-based solution that immediately restricts the prospect of system hackers to spam the system and reduces the demand for setting a predetermined size for new trade blocks.

Ethereum TradingHow can Ethereum trading resemble in training? Given that Ethereum is a people valued usefulness, answering this
question requires well-designed and thought-out principles of data analysis. At the outset, you need to understand that the
current state of the Ethereum undertaking, how the marketplace is growing and the progress that has been made from the core
development team.

Let us begin with the price.

PriceIt is a fact that there is no true value that You can assign to any digital

advantage. The Ethereum trading platform provides
clarity regarding what the traders and users believe is the worth of the Ether. This metric is also claimed to be a sign of the
total confidence in the Ethereum undertaking.

As an investment, Ether has shown Similar development as that of Bitcoin currency. At the time of Ethereum's first audience sale,
users could buy 2000 Ether using 1 BTC and that was trading for just over $600. Since that time, Ether has seen its own price rise
and drop. Of particular importance is the fact that speculators always appear to be attracted to the coordinating activity around
major Ethereum project releases.
Still, such sliding movements have been Trivial compared to Ether's overall price appreciation. At the time of the audience sale,

the price of 1 Ether was approximately $0.30. Compared to the value of $14.30 at the time writing this book, this denotes a 4,666%
rise in worth. An analysis of the Ethereum system's Blockchain proves that company is now pushing the majority of volumes, even
though how much could be defined as hypothetical is uncertain.

Ethereum WalletThe first step towards becoming started with Ethereum is putting up the pocket. The same as a true wallet where you
store coins and notes, an Ethereum wallet enables you to store your Ether. An Ethereum pocket is simply a file that you create for
keeping your Ether--the exact same way you would create a bank account to keep your money.

The wallet you produce could be stored on Different devices or computers (recall, Ethereum is a distributed system). Therefore, you can replicate the wallet file. The pocket file comprises of two chief parts namely: the document--that stores your

Ether--along with the wallet application-- the program that opens the document on your PC.

Now, how can you set up a wallet file on your computer?

Well, you have to have Bitcoin. This is because it is Hard to purchase Ether with a charge card whereas the process of exchanging

Bitcoins with Ether is relatively simple. Simply put, you need an official Ethereum wallet and Bitcoin that comprises some coins

that you begin. Do keep in mind that both wallets must be up-to-date.

What happens when you make an account?

Each account that you produce is characterized by Two keys specifically, the Private Key and the Public Key. At this phase, it is

important to be aware that both the public and private keys are just encryption

algorithms which are utilized to secure your
account whenever you perform any transaction. Both keys are saved as a single file.

Conversion from Bitcoins to Ether

To convert from Bitcoins to Ether, follow the Steps under:

· Log on to Poloniex.com and create an account.

· Log into your account.

· Click on the "BALANCES" tab and choose the "DEPOSITS & WITHDRAWALS" link.

· Utilize the search button to find the Bitcoin's address.

· Copy the Bitcoin address and return to the Coinbase.

· Send the Bitcoins that you would love to invest in Ethereum.

· Await the transaction to be completed.

· Click on the "ETH" marketplace and kind in the sum of Ether which you would like to get.

Ethereum safetyIt is your responsibility to look after your Digital money. In actuality, there are no built-in back apps that will

help restore your Ether if something goes wrong. Because of this, it's vital to maintain your wallet secure, backed up and in sync
with the Blockchain.
Bear in Mind that every pocket which you produce has Two keys: the Private Key and the Public Key. The most important function of
these keys is to aid with the safety of trades which take place online so that even if third party persons intercept them, they
will have a difficult time finding and breaking out what is being sent across the network.
Both Private and Public keys saved as a single file. The very first guideline for establishing a safe Ethereum system is
understanding where the document is stored. In most methods, the key store file would be found in the key store subdirectory of
the Ethereum's data directory. Needless to say, that is dependent upon the type of OS that you're using.

Let's find out just how different OS access this file.

#1: Windows OSIf you are running a Windows OS (any version), Subsequently the node's key store info directory will be found at:

C:\Users\username\%appdata%\Roaming\Ethereum.

You should substitute the username with your name In order for one to access the key store file. For example, if your name is Peter, then you can navigate to C:\Users\peter\%appdata percent\Roaming\Ethereum to find the important file.

#2: Linux OSIf you are running a Linux OS, then you can Get into the key store file in the following location:

Linux: ~/. ethereum

For you to get this document, all you have to Do would be to start the Terminal. Ensure that you've got administrative privileges

to access the file--use the su command and type your root password. After that,

simply type ~/. Ethereum and you're going to
access the key store file.

#3: Mac OSFor Mac OS, the key store will be found in the Following route:

~/Library/Ethereum

Note that the route to the main store file will Always be hidden from users. Therefore, you need to use an appropriate method--that's permitted for your OS--to unhide the files.

Chapter 16: What You Need To Get Started

So, if you're still up to this Chapter, that assumes that you're interested in Ethereum trading and investing into the cryptocurrency platform still. If that's the case, you might be asking yourself "Great. How do I get started?"

Getting started is the easy part, and out of everything in this guide, something you can do right now with nothing more than what you have and no extra cost.

The first thing you'll need to do to get started is to decide how you intend to earn Ethereum Coins. Do you want to perform Ethereum mining, using your computer and it's resources to compete with other miners for parts of a coin before being able to fully put it together? Do you want to just start offering services through the internet and sell your skills and products to earn them? Or do you have disposable income in more traditional currencies that you want to use to purchase them through the internet instead and simply hold onto them? Do you want to do a combination of both of them?

The answer really depends on your intentions, because each decision requires different resources.

With mining, it's all dependent on your computer and it's CPU power. Naturally, weaker and older computers will continually be lagging behind in terms of power and might not be able to mine efficiently enough to justify the cost and time you're putting into mining. The programs that you'll be tasked with running aren't just something you put in the background of your computer and go about your day. They require the full power of every available resource your computer has and will be running them at full capacity. To a lot of Ethereum coin miners, they generally buy one or two very powerful CPU's and use them both as dedicated mining machines, built for no other purposes. Sometimes they'll pool their collective resources together and join what's called a "pool" where other people link up their computers and perform mining tasks together, doling out coins equally among all other members. Other miners buy server space from other people and use the server itself as a tool for mining if they don't have the funds or time to dedicate to a mining computer rig. In a lot of ways, unless you have the computer already designed for mining and time and dedication to mining itself, the cost of mining can quickly outweigh what you'd get in return. Sometimes the effort isn't worth it, but that all does depend on other factors.

In every other case, all you really need is access to the internet and a decent computer on the low to midrange. Most of what else you need only takes up a certain amount of information and can be accessed much like any other internet application.

With that said, the next thing you need to focus on is what sort of wallet system you want to use.

Wallets

Now, what is a wallet? It's been mentioned before in this e-book, but not touched on too much. Is it like a traditional wallet that's used to store money in it?

Yes, and no.

Traditional wallets are generally just that. They hold your finances easily in one place, as well as other information. Digital wallets are used much in the same way but are heavily specialized in holding cryptocurrencies. In a lot of ways they're similar to the kinds of different wallets you would fine offline as well, meaning that different wallets that you find will do different functions.

Some wallet programs are designed to be specialized, going for more specialization purposes than anything else, usually focused on security and information holding. These are perfect for someone who only want to focus on one cryptocurrency at a time, and to make sure that what they earn is safeguarded against people who would want to get their information and hack into their accounts to steal their money. While a lot of their security measures are pretty standard, a lot of times these sort of specialized wallets offer services offline as well such as customer service numbers you could call, IP tracking, printable forms, and such like that.

On the other end of the spectrum though you have wallets that are designed to be more universal. Often these wallets will have a myriad of different options for holding different kinds of cryptocurrencies, not just Ethereum coins. These are usually used for people who are more broad in their investments and are especially useful for those that do eventually want to invest in more than Ethereum. A lot of universal wallets have room inside for all the different cryptocurrencies in place, even for ones designed as a joke. One of the best things about these types of wallets is that you can even use them to transact with other people who use a similar program, exchanging one type of cryptocoin for another. It's not uncommon to see people trading less valuable coins for Ethereum or other high valued coins.

For the purpose of starting out, however, it's often safe to just start with the wallet program that comes with Ethereum on their website. Not only is this wallet highly specialized and focused on the Ethereum coin itself, but it's also tied in with the company that developed the altcoin so you'll have no problem troubleshooting if any problems arise. It's highly recommended that if you do seek out other wallet programs out there afterward, though, to do your research beforehand so that way you don't end up scammed, and having your hard earned investment taken from you.

With that said though the next thing you have to focus on is your internet access, and securing your computer from outside intrusion. This goes without saying, though, as it's often a smart idea to provide protection for your computer, even if you're not involved in the Ethereum trading business, but by reminding yourself to keep yourself protected you'll run less of a chance in the future of having your coins stolen from you. Simple firewall extensions, rootkit tools, anti-virus, and even anti-malware tools will help ensure that no one has access to your information via backdoor programs. Always make sure to backup any valuable information, including wallet information, pool information, or even business information and be vigilant.

Chapter 17: Ethereum's Future

While the future of Ethereum is looking extremely bright, the future of mining on the Ethereum platform is much less black and white. This is because Ethereum has announced plans to switch from a proof of work model to a proof of stake model at some point in 2018. Unlike the proof of work model, the proof of stake verification system will theoretically decrease verification times and deal with issues that are inherent with the proof of work model at a larger scale. This update is in line with a previously posted timeline outlining Ethereum's plans to remain innovative in the blockchain design space.

In May 2017, the platform released the implementation guide for a hybrid proof of work/proof of stake system that is going to be rolled out to test the new system before it becomes the blockchain's primary verification system. The current plan states that the blockchain will alternate between the two systems in such a way that one out of every 100 blocks will use the new system while the rest will continue using the old for the time being.

The hope is that the new system will improve the rate at which new blocks can be produced, which marks the first step in Buterin's continued plans for Ethereum's evolution. When the system goes wide it will mark the first time a proof of stake system has been used to secure a blockchain, which will be a major step forward, despite the modest initial rollout. It will serve as the proof of concept test for an alternate to the proof of work model that has dominated the early days of cryptocurrency development and thus provide proponents a chance to finally test their claims of its superiority. One thing that is already known for sure is the fact that when it is eventually rolled out on a larger scale the proof of stake model will reduce the amount of electricity required to verify a block significantly.

To understand why this change could be so huge for Ethereum, it is important to understand just how it differs from the proof of work model. With a proof of stake verification system, instead of having the miner solve the equation in order to verify the block, a validator, who is confirmed reliable by the stake they have in the system, will commit to its accuracy, knowing that if they lie they will lose their own ether as well. The Alliance is currently testing the new system through a limited use verification process to make sure it is ready for a wider launch next year.

During the first stage of deployment, all of the blocks that are verified through the new system will also be verified through the old system as well. This will help to double verify the blocks contain the information that they should while also testing the accuracy of the new system at the same time. It will also mark the first time the new fork rules will come into play that users have an actual choice about. Validators will then look at the various chains that are available and then make a decision based partially on how much ether is currently in the chain. If they choose poorly, they will then lose money.

This process with then form a consensus that leads to a single larger chain from many smaller ones.

The completed smart contract with then be added to the blockchain proper where it will form what will be known as a Casper account that anyone who is interested in becoming a validator can sign up for. They will then need to deposit a set amount into the system in order to keep them honest. With this done, they will then be able to take part in the new virtual mining process.

Assuming this stage goes smoothly, the second stage will be to deploy it on a much larger scale. This phase will also involve the creation of what is known as a difficulty timebomb. This malicious code will work to make mining the Ethereum blockchain via traditional means more and more difficult over time. The idea being that these spikes in difficulty will make it more trouble to mine than it's worth, while also decreasing the effectiveness of decentralized apps that work inside the old system. Eventually things will get so bad that even the diehard holdouts will have no choice but to switch to the new system.

This doesn't mean the transaction will still not have its issues, however, the first of which will be the chance that the change causes a hard fork instead of a soft fork. If this occurs, and enough people refuse to switch to the new system no matter what, then they could get together and fork the Ethereum blockchain in such a way that it mimics the Ethereum blockchain before the timebomb was put into effect, thus creating a third competing type of Ethereum besides the primary chain and Ethereum Classic which was created as the result of a hard fork that was caused after a large number of ether was stolen.

While not without its hardships, that doesn't mean there aren't several other advantages when it comes to the proof of stake model as well. The biggest of these is the fact that switching to a virtual mining process would decrease the more than 1 million dollars per day that Ethereum miners are currently spending on electricity. This works out to roughly 36 million dollars a month and more than half a billion dollars each year. If the new proof takes off then those costs would be reduced by an estimated 90 percent.

Furthermore, it will also make the mining process more egalitarian as it will no longer be based around who has the most state of the art mining machine, all mining will be handled virtually by the blockchain itself. Even better, it will also make 51 percent attacks much more difficult to complete successfully. A 51 percent attack occurs when a group of miners band together to control more than 51 percent of all nodes running a particular blockchain in an effort to add completely false blocks to the system that the no affected nodes will then accept as true because a majority of the nodes are already reporting it that way.

It also takes things another step further by ensuring that all validators will automatically have a vested interest in staying honest as any foul play will forfeit the funds they added to their Casper account in the first place. Finally, it also makes the process of creating new blocks much faster while also improving scalability by bringing the idea of sharding to blockchain systems. The proof of stake model naturally increases scalability as it then becomes much less complicated to determine the authenticity of blocks as all they will require is knowing who has the greatest stake and who else has the most hashing power. Thus, coming to a consensus can happen almost instantly.

Sharding is the process of segmenting a large database into a set number of manageable shards. Doing so will allow each shard to have its own validators who then would complete their own transactions within their own shard which is why the new model is considered a prerequisite for the process. Once sharding has taken place, it will then help with scalability in a specific way. When it comes to the proof or work model, only so many miners can work on the only available proof at a given time. By instead separating all of the validators into different shards, you allow them all to work on different problems at the same time, increasing the overall speed of the system as a whole significantly.

It's not all going to be good news however, as the new system has some downsides as well. First and foremost, while the odds that the new system is going to work as intended is high, it is by no means guaranteed. There is a possibility that the primary blockchain could be harmed if the transactions aren't processed as planned or if a smart contract is written incorrectly. To combat this possibility, the Ethereum team is working hard on what is being called the finality property which is going to ensure that the current state of the blockchain is secure before the new one can be brought online.

Chapter 18: Ethereum And Bitcoin –

A Comparison

Most of the people reading this book have already heard of Bitcoin. The crypto-currency that was created by Satoshi Nakamoto brought a significant change in the way the world views digital money. Not only has Bitcoin been more than successful in the past decade, but it has also become the most dominantly-used currency in the crypto-currency market. Though this book is on Ethereum, it is useful to compare the two and see how similar or different they are.

There are quite a few differences between the two crypto-currencies, despite one having a broader network.

While Bitcoin was created specifically to become a means for peer-to-peer exchange of digital currency, Ethereum was made to be more than just a crypto-currency. Ethereum provides convenient ways to pay for items online, but it does not stop there. It expands further by giving developers the opportunity to build apps on the Ethereum platform. The platform was created to make it easier for programmers to build decentralized apps without leaving the security of the protocol. The protocol is a set of pre-existing rules that govern how information or data is shared between two or more electronic devices such as computers.

The many different aspects of Ethereum include the Ethereum Virtual Machine, Smart Contracts and the digital currency itself, called Ether. Therefore, Ether is just one of the many uses of the Ethereum platform. This shows that Bitcoin and Ethereum have different intentions in what they aim to achieve and can, therefore, be termed as different projects. Their purpose is the ultimate distinction between the two. Bitcoin was created as an application of the blockchain network. Its purpose is to allow the secure exchange of digital currency between peers. Ethereum offers more applications of the blockchain system aside from having a digital currency. Most popular are the

Smart Contracts that we discussed earlier on in the book.

The following is a detailed comparison of Ethereum to Bitcoin. This comparison will aim to show the differences in various aspects of the two platforms:

Block Time

In Ethereum the block time is often aimed to last between 12-15 seconds. This is very little time compared to the block-time on Bitcoin which lasts 10 minutes. Ethereum is capable of such a fast block-time because of the use of its 'Ghost Protocol'. The Ghost Protocol is used to speed up the time it takes for a transaction to go through and become verified. It also reduces the inevitable greed of some miners creating centralization by 'pool mining'. Pool mining is just pulling your resources together to get rewarded while cutting out anyone who is not in your pool.

Flexibility

As mentioned earlier in the book, Ethereum has a Turing Complete code. This internal system is responsible for the completion of all the algorithms on the network, if it is given enough time and computation power. This means that anything can be calculated on the Ethereum platform. The same cannot be said for Bitcoin, as it does not have the same flexibility due to lacking the Turing Complete aspect. Algorithms that can be solved on Bitcoin are limited to a particular set of rules or programming languages.

Ownership of Tokens

The available tokens on the Ethereum network will be half-mined by the fifth year of its existence. The tokens available on Bitcoin are already more than two-thirds mined. The majority of Bitcoin tokens are owned by the early miners on the platform.

The reason for this is because, when Bitcoin was released, all of the first miners got a bigger portion of the tokens. The Ethereum platform was crowdfunded, and a pre-sale of tokens was made in exchange for Bitcoins.

The Method for Costing Transactions

The two platforms have different methods for charging operations. The transactions on Bitcoin are carried out similarly and compete equally with each other. The transactions on the Ethereum network, on the other hand, are costed on a different basis. The costing on Ethereum is made based on the complexity of the transaction, the storage needs of the trade and the bandwidth of the transaction. The transactions on the Bitcoin platform are also not subject to the size or computational power required to carry out the operation. Furthermore, the Bitcoin transactions are also limited by the scale of the block.

Economic Model

The economic model of the Bitcoin platform is not the same as that of Ethereum. The Ethereum platforms release the same amount of Ether each year. The number of Bitcoin block rewards halves every four years.

Mining Methods

The Ghost Protocol on Ethereum discourages centralized pool mining. If a person is part of a pool, it will bring no advantage to them regarding block propagation. This means that they can only gain by solving a problem on the network and getting a reward.

On Bitcoin, centralized pool mining is allowed. This is how the early miners got to snatch up most of the tokens. It's like having a group of friends who work together to beat the other people in solving a problem. You can then start solving another problem and have a head start. In this case, the head start is the propagation of the next blockchain, which results in rewards in the form of digital currency.

Hashing Algorithm

Ethereum has the hard-hashing algorithm that discourages the use of ASICS like those utilized in the mining of Bitcoin. ASIC is an 'application-specific integrated circuit'. Ethereum instead encourages the use of decentralized mining. This form of mining by individuals using their general processing units is what Ethereum aims to achieve.

Chapter 19: How To Buy, Sell, And Store Ether

This is probably the most interesting part of the book as this chapter will teach you how you can actually make a profit by buying and selling ether. Now, before you can start buying and selling ether for profit, you first need to have a place where you can store ether. This is referred to as an Ethereum/ether wallet. There are several kinds of wallets that you can use. For starters, there are two main kinds. They are the hot and cold wallets. A hot wallet is the kind of cryptocurrency wallet (also applies in other cryptocurrencies and not just for ether) that is stored and held completely online. There is also what is known as a cold wallet, also known as cold storage. This is the kind of wallet where you keep your public and private keys offline. So, how are they different from each other? Well, let us just say that a hot wallet is more convenient to use since all you need is an Internet connection to access and manage this kind of wallet. However, since a hot wallet is exposed to the hazards of the Internet, it is exposed to risks. It can be a target of hackers and even be attacked by viruses and malware. This is where a cold wallet comes into the picture. A cold wallet is not exposed to the Internet; therefore, it offers more security. However, it is not as convenient to use as a hot wallet. Now that you know the two main types of cryptocurrency wallets let us quickly go through their specific types. Hot and cold wallets are further divided into the following:

☐ Web wallet

This is a kind of hot wallet that is completely accessible and manageable online. Hence, it is also referred to as an online wallet. This is the most common type of wallet used by a majority of Ethereum and cryptocurrency users. If you are just starting out, it is strongly advised that you try opening an Ethereum web wallet. A good place, to begin with, is Coinbase since it will allow you to buy and sell ether directly from your Coinbase wallet.

☐ Mobile wallet

A mobile wallet is a hot wallet that you can download on your mobile phone. Most web wallets are also mobile wallets. Normally, you can download the wallet as an application for free from the GooglePlay and/or Apple store.

☐ Desktop wallet

A desktop wallet is kind of cold wallet. When you use a desktop wallet, you store your private and public keys on a computer. The computer does not necessarily have to be a desktop-type computer; hence, a laptop computer will work just fine. Just be sure that it is free from viruses and malware, and do not connect it to the Internet once you start using it as a cold wallet.

☐ Hardware wallet

A hardware wallet is another type of cold wallet. It also works just like a desktop wallet; however, this time you will store your public and private keys on a hardware, such as a USB or a hardware that is specially made for this purpose.

☐ Paper wallet

A paper wallet is a popular type of cold wallet. When you use this wallet, you will store your public and private keys on a paper. Normally, you will be asked to print some codes on a paper. It is also common to include a QR code that you can scan. Without the paper, you will not be able to access and/or move cryptocurrency from your account to another. Hence, the paper is required for any transaction to be made. Needless to say, you should keep it in a safe place. It is advised that you keep several copies just in case you might lose a copy.

Additional tip: Trading Broker Account as a Wallet

Many investors and traders simply use the trading account provided by their broker as a wallet. After all, it is also a place where you can store ether and even other cryptocurrencies. However, if you simply want to invest in Ethereum for a long-term, then you may no longer need a trading broker as you can usually purchase and also sell ether directly from a wallet such as Coinbase. So, if you just want to deal with Ethereum alone, then you may skip the need of working with a trading broker; however, if you want to trade ether as well as other cryptocurrencies, then signing up for an account with a cryptocurrency trading broker like Bitfinex and binance is important. You may also want to have a trading account in cases where you cannot purchase ether or any other cryptocurrency directly from your wallet. There are traders like eToro that will allow you to buy ether in exchange for fiat money, while others will first require you to deposit bitcoins and have them exchanged for ether on the platform. This may vary from trader to trader, so be sure to learn and understand as much as you can about the trading broker. Be sure to read the terms and conditions.

Cryptocurrency Trading Broker

Here are some of the things that you should look at choosing a trading broker. Keep in mind that if you want to buy and sell ether and other cryptocurrencies, it is important that you work only with a trustworthy broker:

☐ Latest reviews

Of course, just like before you use any service or products that are offered online, the first step is always to check the latest reviews and see what other people say about it. The same applies when choosing a trading broker. Be sure to read the reviews and take note of the dates when the reviews were written.

☐ Customer Support

The broker has to have an active and professional customer support team. You will find this helpful, especially if you encounter some technical problems in the future. Normally, a broker will provide you with an email address where you can contact the support team. It may also provide an on-page live chat service or even a number that you may call for inquiries.

☐ Withdrawal requirements

It is usually fast and easy when making a deposit; however, problems may arise when making a withdrawal. It is common for trading brokers to request for certain documents to be submitted prior to processing a withdrawal request. They may require you to submit a copy of a valid ID and a proof of billing. Before you make any deposit, you should make sure that you have such documents available in your possession to avoid problems in the future. Feel free to contact the support team to be clear about this matter. You should also check how many days it will take your broker to complete a withdrawal request. Ideally, a broker should be able to send to you your requested withdrawal within 24 hours, provided that you have already submitted the required documents.

☐ Trading platform

Your broker is the one who will provide you with a platform for trading cryptocurrencies. Since you want to invest in ether, then the platform should make it easy for you to buy and sell ether at any time that you want. It should also provide you with free tools, such as price charts, to help you understand how ether is performing in the market. This is essential, especially if you are fond of using technical analysis.

☐ Mobile feature

These days, it is much easier to access the Internet through a mobile device. Not only should your broker provide you with a trading platform, but it should also allow you to buy and sell ether directly from your mobile phone. Do not worry; most, if not all, reliable trading brokers always have a mobile version of the trading platform.

Buying and Selling Ether Tips and Tricks

It is now time to discuss the best practices that you should observe when you buy and sell ether for profit. Pay close attention to these tips and tricks to increase your chances of success.

☐ Buy low, sell high

This is the most common advice given to traders. The same applies when you invest in ether: Buy low, sell high. Take note that this does not mean that you have to wait for the price of ether to drop down back to

$2 for this will most likely not happen anymore. Buying low simply means buying it at a price that is lower when you get to sell it. Hence, although the price of a single ether these days is worth more than a thousand dollars, it is still "low" provided that you can sell it at a higher price. So, the important thing to note here is to identify if the price of ether will most likely increase. If yes, then you should make a buy order. Of course, for you to be able to predict the price movement of ether in the market, you cannot just rely on luck or mere guesswork. You need to research and analyze the market. When it comes to understanding how ether performs in the cryptocurrency market, you should learn two important strategies: fundamental analysis and technical analysis.

☐ Fundamental Analysis

Fundamental analysis is rightly called the lifeblood of investment for good reasons. This approach deals with the fundamentals which means that it deals directly with the basics. Predicting the price movement of ether or any other cryptocurrency mainly depends on how well you understand the basic factors or elements that have an influence over it price, such as the economy, the level of competition among the different cryptocurrencies in the market, the latest news, market acceptance, government regulations, and technological developments and breakthroughs, among others. When you use fundamental analysis, the key is to gather as much quality information as you can. As the saying goes, "Knowledge is power." The more knowledge that you have not only about ether but also about the whole cryptocurrency market itself, the more likely that you can "read" and predict the direction that ether will most likely take in the market. When you use this approach, it is important for you to pay attention to the news as it can strongly affect ether and other cryptocurrencies. For example, when a positive news piece about Ethereum was featured on Google and CNN, the price of Ether surged even further. If you are serious about making continuous profit, then fundamental analysis should be part of your day-to-day activity as a cryptocurrency investor.

☐ Technical Analysis

If there is fundamental analysis, then there is also technical analysis. Technical analysis is where you study the price movements (past and current trend) of a particular cryptocurrency as shown by graphs or charts. This is an excellent strategy if you are more of a visual person. Indeed, it is also much simpler than fundamental analysis which is why so many investors like this approach. The idea behind this strategy is that all of the factors that can affect a cryptocurrency (in this case, ether) have their final effect on the price. Therefore, by simply analyzing the price of ether, you also get to deal and analyze all the factors that have an influence on it. If you like taking advantage of patterns and trends, then this is the one for you. It is worth noting, however, that technical analysis alone may not always be enough. Many experts agree that to further increase the effectiveness of this strategy, you should combine it fundamental analysis, and that mastery of the two can increase your chances of making the right investment decision by more than 75%.

☐ Research

Research is definitely in the heart of every successful investment. This is a very important part of fundamental analysis. If you want to improve your chances of success significantly, then you should do your research. A common mistake is not doing continuous research. Keep in mind that ether has a high volatility, especially these days when it is constantly experiencing high surges in price. You need to be up-to-date with the latest news and developments. Another common mistake is not doing enough research. It is true that many investors do their research prior to making an investment; however, the problem is that many of them do not render sufficient research. Just because you have studied a particular graph for about three hours does not mean that you are already in the position to make a sound investment decision. You should know that professional and successful cryptocurrency investors do such studies and research on a regular basis and yet they are still very careful every time they make an investment. Be very keen on doing research. The more that you research and understand ether the more easily you can predict its price movement.

☐ Buy & Hold

This is probably the simplest and very effective strategy that you would love to learn when you invest in ether. Indeed, many real-life success stories are based on this strategy. Its power lies in its simplicity. As the name suggests, it is about buying ether and then holding on to it for some time. The idea here is to wait for its price to increase. You can then sell it at a profit at some future date. Take note that you cannot just apply this strategy at any time that you want. The proper way of using this approach is to first study the cryptocurrency concerned, ether. You should apply the buy and hold strategy only if it appears that the price of ether is most likely going to increase. Hence, when you think that the price is most likely going to surge higher and higher, then it is time to use this strategy. However, if after doing your research, you realize that the price of ether is most likely going to drop, then do not use this strategy. Another important part to take note of is when to sell your ether at a profit. Do not underestimate how highly volatile ether is. Sometimes holding on to your ether for too long may not be a good idea. To be sure, you need to keep a close eye on the market and continue to study Ethereum, so that you will know if you should still continue to hold on to your investment or if it time for you to sell it at a profit. Also, before you sell your ether, remember that there is a difference between the buy price and the sell price. The buy price is always higher than the sell price. This is also how a trading broker makes money. Consider the difference and make sure that you sell at a profit.

☐ Only invest the money that you can afford to lose

A common advice that is given to casino gamblers is to gamble only with the money that they can afford to lose. Although investing in ether is not gambling, the same advice still applies. After all, no matter how much you research and study the market, you can only increase your chances of making the right investment decision, but you cannot guarantee the return of positive profit. Hence, you should only invest the money that you can afford to lose. Just like any other investment, investing in ether or any other cryptocurrency has its risks. Therefore, do not invest using the money that you need to pay for your household bills and other obligations. This is also a good way to prevent you from trading under pressure. When you are pressured, your emotions may cloud your judgment which will prevent you from thinking objectively. When you deal with the cryptocurrency market, it is important that you can think clearly and decide objectively.

☐ Join online groups and forums

You should join and participate in online groups and forums on cryptocurrencies. This is a good way for you not only to meet people with a similar interest but you can also learn interesting views and strategies from them. Many cryptocurrency developers are also active in such places, so this is a good way for you to gain valuable information.

☐ Learn about other cryptocurrencies

Although your main interest is investing in Ethereum, it is still important for you to keep an open mind and learn about other cryptocurrencies. Do not forget that Ethereum is not the only profitable cryptocurrency in the market. Also, these other cryptocurrencies also affect how ether performs in the market since they are in competition with one another. By studying and learning about other cryptocurrencies, you will be more able to understand why ether behaves in a certain way.

☐ Learn to wait

Sometimes the market can be difficult to predict despite the amount of research that you do. Indeed, even in the past, the price of Ethereum also experienced some instability. Although it has already gained popularity and established itself in the market today, it does not mean that it will no longer face any problem. A common mistake committed by many investors and traders is to enter a position even when they are not confident enough if it would turn out to be profitable or not. You should keep in mind that you should only invest in ether if you are convinced based on your research that its price is most likely going to increase. If you are not that confident that it will be profitable, then learn to just wait it out. It is common for ether to be hard to predict. However, there are also moments when you are almost 100% certain of its price movement. You can then take advantage of such momentous occasions to make a profit. Be patient and always observe proper timing. Do not forget that you are not in any way compelled to make any investment. But, when you see that ether is getting the best of the market, then be sure that you are there to take full advantage of it.

☐ Have fun

Enjoy it. Enjoy the perks of being an investor. You can also think more clearly and be a more effective investor/trader if you are having fun. Of course, do not confuse this with making reckless decisions and investments. To have fun means to enjoy the process and the life of being a cryptocurrency investor. Hence, enjoy the process of doing research on Ethereum, as well as other cryptocurrencies. If you want, you might want to team up with a friend or simply another investor who is interested in Ethereum. Also, do not take it too seriously. Sometimes being too serious can hinder your creative mind from harnessing its power and coming up with interesting and profitable ideas. So, enjoy and have fun, yet stay professional.

Chapter 20: Useful Tips

Secure your password

Secure your Ethereum account with a unique password just as you would with your social media account and bank account. You don't want to compromise your money. Accounts with a simple and patterned password are more accessible to hackers. For example, one of the common passwords many people use is their birth date using numbers that correspond to the month, date, and year. DO NOT USE SIMPLE PASSWORDS. Use unique ones. A difficult password will keep attackers out of your account. A secure password contains letters, numbers, symbols, and must be at least 10 characters long.

The longer the password, the harder it is for attackers to know it. Of course, if you want to use a long password, you should remember what it is. You might want to write it down in a secret paper, or at least write down hints that will make you remember your password.

Your wallet must be encrypted

Cryptocurrency wallets are sensitive. They store information about the funds of your digital currency. There are many wallets available today. You must choose the one that has high protection level for the users. You can select the trusted wallets like MyEtherWallet.

Backup your wallet

Having a back-up for your wallet offers several benefits. With your wallet backed up, you can save your account in case there would be system failures.

Chapter 21: Top Tips To Make Your Ethereum Experience Better

If you're not too interested in investing in the future of Ethereum technology, and are more interested in using Ethereum correctly, then you will probably want to know about how you can optimize your Ethereum experience. That is what this chapter will cover. This chapter will get into some of the nuances that exist within the Ethereum application that can take your experience to the next level. If you can avoid mistakes that others have made before you, why wouldn't you seek to do so?

Top Tip 1 for Ethereum Users: Recognize Ethereum's Shortcomings

One of the first tips from which any Ethereum user can benefit is to recognize the current limitations of Ethereum. Sure, the potential appears to be on the horizon for Ethereum and its Smart Contracts, but it still has a long way to go before this application should be considered foolproof and completely legitimate. Some of the current problems surrounding Ethereum include the following: scaling to meet demand, figuring out how to audit people properly, and how to properly manage an increasingly growing network. If you approach Ethereum as if everything within its digital borders has been completely figured out, you're simply not looking at the technology realistically and will run the risk of making mistakes with your ether based on those assumptions.

Top Tip 2 for Ethereum Users: Know Some Programming

If you're a n00b when it comes to programming, it might not be a good idea to jump into using Ethereum blindly. Before you start using Ethereum, it's recommended that you at least have a basic understanding of how Javascript works. If this does not seem like an immediate possibility for yourself, it might be a good idea to at the very least find a friend who you trust and can walk you through the basics of Ethereum or work on the network on your behalf.

Top Tip 3 for Ethereum Users: Keep Yourself Educated

This cannot be overstated. When thinking about Ethereum as a global technology that can connect people internationally through decentralization, there are multiple facets of society that need to be working together. For example, complete implementation of Ethereum in society will require not just the storage and technical capabilities from thousands of computers to be used as mining stations; it will also require much more. Most notably, the question of how governments will deal with the infrastructure costs needed to make Ethereum viable on a national and global scale is one that is still unanswered. Keeping track of how governments of all shapes and sizes are responding to Ethereum and other blockchain technologies in general will allow you to better understand where Ethereum is headed and how you fit into this larger networking scheme.

Top Tip 4 for Ethereum Users: Always Check Before You Send

Whenever you're sending ether to a new party, you're going to want to make sure that everything seems to be valid prior to sending large amounts of currency. To do this, simply send over a small amount of ether to whomever is receiving it, prior to sending any large amount. This will help to ensure that you're not being duped, and that everything is checking out as it should be. Sure, Ethereum appears to be safer than Bitcoin, but that does not mean that shady characters do not exist on the Ethereum network. You never know who the other person is at the end of your transaction, and this is why it's incredibly important to keep your money as safe as possible.

Top Tip 5 for Ethereum Users: Guard Your Private Key with Your Life

While your public key can be seen by anyone on the Ethereum network, your private key is what fully protects the ether in your wallet. It shouldn't have to be stated, but you'd be surprised at how many people are rather careless with their private key. When you think about the private key as being the secret code that opens up the safe that contains your precious ether, it should be obvious that you need to be extremely careful with it. Some common things that people have been known to do that have led to the infiltration and theft of their Ethereum currency include emailing their private key to someone, posting their private key on social media accounts such as Twitter or Reddit, and storing their private key in their Dropbox. Your email and your Dropbox can be hacked more easily than you think, often through simple phishing techniques, and there is simply no excuse for posting your private key to places where others can see it freely.

Top Tip 6 for Ethereum Users: Make Sure Your Work is Correct

Another simple tip that could end up saving your money is to check your work before sending anything. Similar to when you're transferring money to another account on your computer or posting money to someone on Venmo, an added zero could be the difference between you sending someone ten dollars and one hundred dollars. Within Ethereum, this is even more critical. Sending denominations of currency in decimalized amounts will likely take some getting used to. Along these same lines, it's important to check that the address to which you're sending your currency is correct prior to sending it. It would be a shame to send the wrong person your currency without trying to do so on purpose.

For the tips that were presented in this chapter related to how to use the Ethereum application, the overarching message is to be careful. Beyond the security that you need to put in place for yourself, keeping up with current events that are occuring within and around Ethereum are also incredibly important for multiple reasons. You don't want to invest your dollars into cybercurrency that is going to end up being a dud. For this reason, keep yourself receptive to constantly learning about how Ethereum is evolving. Lastly, remember that this technology is not yet completely developed. For this reason, you will need to steady yourself against setbacks that may come your way, and be as resiliant as possible.

Chapter 22: The Future Of Ethereum

After going through a lot of battles like DAO hack and the resulting hard fork, Ethereum is a lot more matured now. It can now fully expand and turn into something that is truly unique. So what does the future hold for this cryptocurrency?

From Proof-of-Work to Proof-of-Stake

As discussed in a previous chapter, Ethereum will soon be shifting to proof-of-stake coming from its current format which is proof-of-work. Let's have a general discussion of the two.

Proof of Work (POW) – This is the same protocol being followed by Bitcoin wherein miners extract cryptocurrencies by using computer power in solving complex mathematical equations.

Proof of Stake (POS) – In this protocol, the mining process will become virtual because validators will soon replace miners. Basically, the validator will lock up some of his ether as an investment. He then starts validating the blocks and if he sees a block that can be linked to the blockchain, he can validate by putting a bet. If and when the block gets linked, he gets a reward which is proportional to the investment. If the bet is placed on the wrong block, the investment will be lost. In implementing this new protocol, Ethereum will be using the Casper algorithm. Initially, it will be a combination of both proof-of-work and proof-of-stake wherein most transactions will be done through POW while POS kicks in every 100th transaction. This is to test the effectiveness of POS.

Advantages of Proof of Stake

- Lowers the monetary and energy cost. Cryptocurrency miners spend approximately $50,000 on electricity every hour. In a year, that would amount to $450 million. That's power and money being wasted. Proof of stake makes the mining process virtual which potentially cuts the overall cost.

- No computer power advantage. Doing it virtually means profit will not depend on computer power alone.

- Makes it harder for 51% attacks. When a team of miners gains above 50% of the available hashing power in the world, they can do a 51% attack. This is not possible with proof of stake.
- Validators will be malicious-free. Using malicious or wrong blocks in the chain will result in lost investment so validators will avoid them.
- Creation of blocks. Newer blocks will be created faster.
- It will be scalable. The sharding concept will make the blockchain more scalable.

The Future of Proof of Stake

Casper stage one is going to be implemented wherein proof-of-stake will be used to check every 100th block. Casper scripts are being run through bug detectors to ensure the system is free from mathematical errors.

Eventually, the majority of block creation will be done through proof of stake. First, Ethereum will make mining more difficult exponentially. This reduces the hash rate which in turn reduces the entire blockchain's speed and all the applications running. This forces people involved to switch to proof of stake.

The Disappearance of Ethereum

The coming updates and changes in Ethereum ensure its future as one of the top digital currencies in circulation. The ultimate goal, however, is for Ethereum to blend in and disappear, becoming omnipresent in all transactions. People will eventually be using the technology without even realizing it. There's still a long way to go before this can be achieved and stumbling blocks will be encountered but at its current stage of maturity and resiliency, Ethereum is in a position to do it.

Conclusion

Now you will want to make an Ethereum account and start investing with Ethereum. As you have seen in this book, there are a lot of good things that will come from using Ethereum; especially since it will be the platform that will change the future of the financial sector.

There is no guarantee that the financial sector will be changed because of Ethereum, but there is not going to be the promise that Etherem will be around for an extended period of time. However, as Ethereum has proven, it seems like it will be around because it is continuing to evolve and adapted to the new technology.

You will find that investing with Ethereum will be a lot different than investing with bitcoin. Investing with Ethereum will give you more options for investing, and you will be able to do more with Ethereum.

The smart contracts that you will be able to work will make it so that you can get rid of judges and lawyers that way you can save more money on top of the money that you are already saving because you do not have to rent a building and get office supplies.

In the end, investing with Ethereum will be your choice, but if it was up to us, we would recommend that you invest in it!

Thank you and good luck!

www.ingramcontent.com/pod-product-compliance
Lightning Source LLC
LaVergne TN
LVHW011937070526
838202LV00054B/4687